"This beautifully written and illustrated book elucidates both the mechanics and mechanisms of plastic deformation of metals and describes the testing methods used to reveal those properties.

While their treatment is comprehensive, Clyne and Campbell have kept the focus on essential topics and have produced a highly readable book that promises to be an instant classic.

The book is both authoritative and tutorial and will be welcomed by students and practicing engineers alike."

William D. Nix
Stanford University

"This is the best book on metal plasticity I have ever seen. It is the most authoritative text available, and yet the most understandable. There is sufficient detail for the serious practitioner, but it is also accessible enough for the novice. The coverage of stresses, strains and continuum plasticity through to every conceivable test technique is the most comprehensive available."

Mark Rainforth
The University of Sheffield

"Clyne and Campbell have produced a wonderful book which I recommend thoroughly. It is well-written, easy to read and beautifully produced.

The subject matter is very important, but deceptively challenging. Existing textbooks do not cover it carefully enough. The authors have done a tremendous job to explain the salient points with much-needed rigor and a great deal of style."

Roger C. Reed
University of Oxford

Testing of the Plastic Deformation of Metals

Discover a novel, self-contained approach to an important technical area, providing both theoretical background and practical details. Coverage includes mechanics and physical metallurgy, as well as study of both established and novel procedures such as indentation plastometry. Numerical simulation (FEM modeling) is explored thoroughly, and issues of scale are discussed in depth. Discusses procedures designed to explore plasticity under various conditions, and relates sample responses to deformation mechanisms, including microstructural effects. Features references throughout to industrial processing and component usage conditions, to a wide range of metallic alloys, and to effects of residual stresses, anisotropy and inhomogeneity within samples. A perfect tool for materials scientists, engineers and researchers involved in mechanical testing (of metals), and those involved in the development of novel materials and components.

T. W. Clyne is an Emeritus Professor in the Department of Materials Science and Metallurgy at the University of Cambridge and also the Chief Scientific Officer at Plastometrex Ltd. In addition, he is the Director of DoITPoMS, an educational materials science website, a Fellow of the Royal Academy of Engineering, a Helmholtz International Fellow and a Leverhulme Emeritus Fellow.

J. E. Campbell (PhD University of Cambridge) is Chief Engineer at Plastometrex Ltd. with expertise in mechanical testing of metals with a focus on indentation and finite element modeling.

Testing of the Plastic Deformation of Metals

T. W. CLYNE
University of Cambridge
Plastometrex Ltd., Cambridge

J. E. CAMPBELL
Plastometrex Ltd., Cambridge

CAMBRIDGE
UNIVERSITY PRESS

University Printing House, Cambridge CB2 8BS, United Kingdom

One Liberty Plaza, 20th Floor, New York, NY 10006, USA

477 Williamstown Road, Port Melbourne, VIC 3207, Australia

314–321, 3rd Floor, Plot 3, Splendor Forum, Jasola District Centre, New Delhi – 110025, India

79 Anson Road, #06–04/06, Singapore 079906

Cambridge University Press is part of the University of Cambridge.

It furthers the University's mission by disseminating knowledge in the pursuit of education, learning, and research at the highest international levels of excellence.

www.cambridge.org
Information on this title: www.cambridge.org/9781108837897
DOI: 10.1017/9781108943369

© Cambridge University Press 2021

This publication is in copyright. Subject to statutory exception and to the provisions of relevant collective licensing agreements, no reproduction of any part may take place without the written permission of Cambridge University Press.

First published 2021

Printed in the United Kingdom by TJ Books Limited, Padstow Cornwall

A catalogue record for this publication is available from the British Library.

ISBN 978-1-108-83789-7 Hardback

Cambridge University Press has no responsibility for the persistence or accuracy of URLs for external or third-party internet websites referred to in this publication and does not guarantee that any content on such websites is, or will remain, accurate or appropriate.

Contents

			page
	Preface		xi
	Nomenclature		xiii
1	**General Introduction**		1
	1.1	Rationale and Scope of the Book	1
	1.2	Structure and Readership of the Book	2
	1.3	Basic Elastic and Plastic Property Ranges	3
2	**Stresses, Strains and Elasticity**		7
	2.1	Stress and Strain as Second Rank Tensors	7
	2.2	Transformation of Axes	9
		2.2.1 Transforming First Rank Tensors (Vectors)	9
		2.2.2 Transforming Second Rank Tensors and Principal Stresses	11
		2.2.3 Use of Mohr's Circle	12
	2.3	Representation of Strain	14
	2.4	Stress–Strain Relationships and Engineering Constants	16
		2.4.1 Stiffness and Compliance Tensors	16
		2.4.2 Relationship to Elastic Constants	17
		2.4.3 Engineering Shear Strains, Shear Modulus and Bulk Modulus	18
		2.4.4 Relationships between Elastic Constants	19
3	**Continuum Plasticity**		21
	3.1	The Onset of Plasticity and Yielding Criteria	21
		3.1.1 Deviatoric (von Mises) and Hydrostatic Stresses and Strains	22
		3.1.2 Yield Envelopes in Stress Space – von Mises and Tresca Criteria	23
		3.1.3 True and Nominal (Engineering) Stresses and Strains	27
	3.2	Constitutive Laws for Progressive Plastic Deformation	28
		3.2.1 Quasi-Static Plasticity – the Ludwik–Hollomon and Voce Laws	28
		3.2.2 The Strain Rate Dependence of Plasticity and the Johnson–Cook Formulation	31

		3.2.3 Constitutive Laws for Superelasticity	32
		3.2.4 Progressive Creep Deformation and the Miller–Norton Law	34
	3.3	Residual Stresses	36
		3.3.1 Origins of Residual Stress	37
		3.3.2 Effects of Residual Stress on Plasticity	40
	References		40
4	**Mechanisms of Plastic Deformation in Metals**		**43**
	4.1	Basic Dislocation Structures and Motions	43
		4.1.1 The Atomic Scale Structure of Metals	43
		4.1.2 Concept of a Dislocation	44
		4.1.3 The Dislocation Line Vector and the Burgers Vector	46
		4.1.4 Screw Dislocations	47
		4.1.5 Force on a Dislocation and Energies Involved	48
	4.2	Further Dislocation Characteristics	50
		4.2.1 Tensile Testing of Single Crystals and Schmid's Law	50
		4.2.2 Dislocation Interactions and Work Hardening Effects	51
		4.2.3 Creation of Dislocations and Types of Dislocation Mobility	53
		4.2.4 Rearrangement of Dislocations and Recrystallization	56
	4.3	Dislocations in Real Materials and Effects of Microstructure	58
		4.3.1 Plastic Deformation of Polycrystals	58
		4.3.2 Solution Strengthening (Substitutional Atoms)	59
		4.3.3 Effects of Interstitial Solute	60
		4.3.4 Precipitation Hardening and Dispersion Strengthening	63
	4.4	Deformation Twinning and Martensitic Phase Transformations	66
		4.4.1 Deformation Twinning	67
		4.4.2 Martensitic Transformations	71
	4.5	Time-Dependent Deformation (Creep)	73
		4.5.1 Background	73
		4.5.2 Coble Creep	75
		4.5.3 Nabarro–Herring Creep	76
		4.5.4 Dislocation Creep	76
		4.5.5 Effects of Microstructure on Creep	77
	References		78
5	**Tensile Testing**		**81**
	5.1	Specimen Shape and Gripping	81
		5.1.1 Testing Standards	81
		5.1.2 Geometrical Issues and Stress Fields during Loading	82
		5.1.3 Issues of Sample Size	85
	5.2	Measurement of Load and Displacement	87
		5.2.1 Creation and Measurement of Load	87
		5.2.2 Displacement Measurement Devices	87

	5.3	Tensile Stress–Strain Curves	88
		5.3.1 Nominal and True Plots	88
		5.3.2 The Onset of Necking and Considère's Construction	89
		5.3.3 Neck Development, UTS, Ductility and Reduction in Area	92
		5.3.4 Load Drops and Formation of Lüders Bands	98
	5.4	Variants of the Tensile Test	100
		5.4.1 Testing of Single Crystals	100
		5.4.2 Biaxial Tensile Testing	100
		5.4.3 High Strain Rate Tensile Testing	101
		5.4.4 Tensile Creep Testing	102
	References		103
6	**Compressive Testing**		107
	6.1	Test Configuration	107
		6.1.1 Sample Geometry, Strain Measurement and Lubrication Issues	107
		6.1.2 Buckling Instabilities	108
		6.1.3 Sample Size and Micropillar Compression	109
	6.2	Compressive Stress–Strain Curves	110
		6.2.1 Nominal and True Stress–Strain Plots	110
		6.2.2 FEM Modeling, Sample/Platen Friction and Barreling Effects	112
	6.3	Tension/Compression Asymmetry and the Bauschinger Effect	115
		6.3.1 Tension/Compression Asymmetry	115
		6.3.2 The Bauschinger Effect	117
	6.4	The Ring Compression Test	119
	References		120
7	**Hardness Testing**		123
	7.1	Concept of a Hardness Number (Obtained by Indentation)	123
	7.2	Indentation Hardness Tests	125
		7.2.1 The Brinell Test	125
		7.2.2 The Rockwell Test	129
		7.2.3 The Vickers Test and Berkovich Indenters	131
		7.2.4 The Knoop Test	137
	7.3	Effects of Sample Condition	138
		7.3.1 Microstructure, Anisotropy and Indentation of Single Crystals	138
		7.3.2 Residual Stresses	140
	7.4	Other Types of Hardness Testing	142
		7.4.1 Rebound Hardness Testing	142
		7.4.2 Scratch Testing and the Mohs Scale	144
	References		145

8 Indentation Plastometry — 148

- 8.1 Introduction to Indentation Plastometry — 148
- 8.2 Experimental Issues — 149
  - 8.2.1 Indenter Shape and Size — 149
  - 8.2.2 Penetration Depth, Plastic Strain Range and Load Requirements — 151
  - 8.2.3 Sample Preparation — 154
  - 8.2.4 Experimentally Measured Outcomes — 157
- 8.3 FEM Simulation Issues — 158
  - 8.3.1 Representation of Plasticity Characteristics — 158
  - 8.3.2 Meshing, Boundary Conditions and Input Data — 159
  - 8.3.3 Characterization of Misfit ("Goodness-of-Fit") — 164
  - 8.3.4 Convergence Algorithms — 164
- 8.4 Range of Indentation Plastometry Usage — 166
  - 8.4.1 Presentation of Results — 166
  - 8.4.2 Effects of Material Anisotropy and Inhomogeneity — 167
  - 8.4.3 Measurement of Residual Stresses — 169
  - 8.4.4 Indentation Creep Plastometry — 173
  - 8.4.5 Indentation Superelastic Plastometry — 178
  - 8.4.6 Commercial Products — 179
- Appendix 8.1 Nelder–Mead Convergence Algorithm — 183
- Appendix 8.2 Distribution of Plastic Work in terms of Strain Range — 185
- References — 186

9 Nanoindentation and Micropillar Compression — 192

- 9.1 General Background — 192
- 9.2 Nanoindentation Equipment — 193
  - 9.2.1 Equipment Design — 193
  - 9.2.2 Nanoindenter Tips — 195
  - 9.2.3 High Temperature Testing and Controlled Atmosphere Operation — 198
- 9.3 Nanoindentation Testing Outcomes — 200
  - 9.3.1 Background — 200
  - 9.3.2 Measurement of Stiffness and Hardness — 201
  - 9.3.3 Characterization of Creep — 203
- 9.4 The Nanoindentation Size Effect — 204
  - 9.4.1 Experimental Observations — 204
  - 9.4.2 Size Effect Mechanisms and "Pop-in" Phenomena — 205
- 9.5 Micropillar Compression Testing — 208
  - 9.5.1 Creation of Micropillar Samples — 208
  - 9.5.2 Test Issues — 210
  - 9.5.3 Test Outcomes and Size Effects — 211
- References — 214

10	**Other Testing Geometries and Conditions**	219
	10.1 Bend and Torsion Testing	219
	10.1.1 Mechanics of Beam Bending	219
	10.1.2 Mechanics of Torsion	221
	10.1.3 Three-point and Four-point Bend Testing	224
	10.1.4 FEM of Four-point Bend Testing	226
	10.1.5 Torsion Tests	228
	10.1.6 Combined Tension–Torsion Tests	230
	10.2 Buckling Failure	231
	10.2.1 Elastic (Euler) Buckling	231
	10.2.2 Plastic Buckling	232
	10.2.3 Brazier Buckling of Thin-Walled Structures during Bending	234
	10.3 Cyclic Loading Tests	235
	10.3.1 Background to Cyclic Loading and Fatigue Failure	235
	10.3.2 Fracture Mechanics and Fast Fracture	236
	10.3.3 Sub-Critical Crack Growth	239
	10.3.4 Stress–Life (S–N) Fatigue Testing	241
	10.4 Testing of the Strain Rate Dependence of Plasticity	243
	10.4.1 High Strain Rate Tensile Testing	245
	10.4.2 Hopkinson Bar and Taylor Tests	247
	10.4.3 Ballistic Indentation Testing	249
	Appendix 10.1 Calculating the Second Moment of Area, I	254
	Appendix 10.2 Beam Deflections from Applied Bending Moments	256
	Appendix 10.3 Mechanics of Springs	261
	Appendix 10.4 Interpretation of Data from Strain Gauge Rosettes	264
	References	265
	Index	270

Preface

The plasticity of metals is central, not only to materials science as a subject, but also to the whole history of technological development. Broadly speaking, the only metal found naturally in elemental form is gold, which has been prized throughout human history – partly due to its lustre, but also because it can readily be formed into various shapes (including very thin foil). Development of the skills and knowledge needed to extract other metals has been transformational for human society and their most important feature is arguably their capacity for permanent shape change, without fracturing. While an understanding of the mechanisms involved is relatively recent, it still extends back several decades. Many thousands of books and papers, published over the past 100 years, cover the topic in detail. This coverage includes both the micro-mechanisms, with the discovery of the dislocation being pivotal, and engineering aspects, usually with the plasticity being treated on a continuum basis.

Several books contain a mixture of physical metallurgy and associated mechanical properties, with mechanical testing procedures often covered in some way. However, relatively few books have been dedicated to testing of metal plasticity. The importance of detailed characterization, and of measurements being made with an understanding of what is taking place inside a sample during a test, has therefore prompted us to produce this book. Its content is based both on several decades of teaching and research experience in Cambridge University and also on extensive interaction with a range of industrial partners and collaborators over that period. This has included, over the past few years, our close involvement with the founding and development of Plastometrex, a company that is oriented towards the development of novel procedures for testing of metal plasticity.

The book has 10 chapters, broadly divided into three sections. The first of these provides background concerning mechanics and microstructural aspects (related to plasticity). The second covers the traditional testing techniques of tensile, compressive and hardness testing, while the third is devoted to various other test configurations and conditions, many involving recent developments. Such testing arrangements include those for study of creep behavior, high strain rates, superelastic deformation, cyclic loading etc., in addition to conventional plasticity characterization. Finite element method (FEM) modeling, which is a powerful tool for investigation of mechanical deformation, figures strongly throughout. The coverage includes "nanoindentation," but is mainly oriented towards obtaining bulk mechanical properties, for which it is not well suited (due to the deformed volume being too small for its response to be representative of the bulk). The important recent advances concerning indentation relate less to it being carried out on a very fine scale, but more to it being instrumented so as to obtain detailed information about the (bulk) mechanical characteristics. This represents a fundamental advance, relative to its origins in hardness testing, and it is explored here in some detail. There is reference throughout to industrial processing and component usage conditions, to a wide range of metallic alloys and to effects of residual stresses, anisotropy and inhomogeneity within samples.

A couple of points may be noted concerning timing. As mentioned above, we were involved in the founding of Plastometrex, which took place in 2018. At around that time, we were starting to feel that a book providing detailed background to conventional test procedures, as well as the potential for novel developments, could have value – nothing substantial of that type was then available. Some exploratory steps were taken during 2019, with serious production of material starting towards the end of that year. Of course, the lockdown associated with the Covid-19 pandemic, which started in the UK in late March of 2020, did affect this procedure, although in fact most of the manuscript preparation had been completed by that time. It did have a dramatic effect on the functioning of the University, although as it happens our involvement there was starting to reduce sharply in any event. Lockdown probably gave a boost to a lot of book-writing, but in our case its main effect was rather one of impeding progress (including the creation of some challenges for Plastometrex). The manuscript was in fact delivered to CUP in June 2020. We would like to thank the production and editing staff at CUP for working closely and helpfully with us during the second half of 2020 – a period during which the lockdown in the UK, while easing in a rather staccato manner, was still causing a range of problems. Of course, none of us know quite how things will pan out during the coming years, concerning the pandemic and other potential sources of disruption. Perhaps one of the few certainties is that metals, and particularly the way that they undergo plastic deformation, will continue to be of prime importance.

Finally, we would like to thank our partners, Gail and Ine, for their invaluable support during the preparation of this book.

Nomenclature

Parameters

A	(m^2)	area
A	(K)	temperature (of austenite formation)
a	(–)	direction cosine
a	(m)	distance between inner and outer loading points (Fig. 10.6)
a	(m)	lattice parameter (cubic system)
a	(Pa)	constant in Basquin law (Eqn. (10.21))
b	(–)	exponent in Basquin law (Eqn. (10.21))
b	(m)	chord length (Fig. 8.6)
b	(m)	Burgers vector
C	(Pa)	stiffness (4th rank tensor)
C	(–)	strain rate sensitivity parameter (Eqn. (3.15))
C	(–)	dimensionless constant in Knoop hardness expression (Eqn. (7.8))
C	(Pa^{-n} s$^{-(m+1)}$)	parameter in Miller–Norton creep law (Eqn. (3.17))
c	(m)	crack length or flaw size
D	(m)	sample thickness or diameter
d	(m)	grain size
E	(Pa)	Young's modulus
e	(–)	relative displacement (2nd rank tensor)
F	(N)	force (load)
G	(J m^{-2})	strain energy release rate
G	(Pa)	shear modulus
g	(–)	goodness of fit parameter (Fig. 10.25))
H	(kgf m^{-2})	hardness number
h	(m)	indenter depth
h	(m)	thickness
h	(–)	parameter relating to Miller indices for crystallographic planes
I	(Pa)	invariants in the secular equation for stress (Eqn. (2.13))
I	(m^4)	second moment of area
K	(Pa)	bulk modulus
K	(Pa)	work hardening coefficient
K	(Pa m^{-1})	constant in Eqn. (10.36)

Nomenclature

K	(Pa m$^{1/2}$)	stress intensity factor
k	(–)	parameter relating to Miller indices for crystallographic planes
k	(–)	dimensionless factor (Eqn. (10.7))
l	(–)	parameter relating to Miller indices for crystallographic planes
L	(m)	length
L	(m)	distance between obstacles (Eqn. (4.14))
L	(m)	spacing between dislocations (Eqn. (4.12))
M	(m N)	bending moment
M	(K)	temperature (of martensite formation)
m	(–)	parameter in Miller–Norton creep law (Eqn. (3.17))
N	(–)	dimensionless number
n	(–)	stress exponent (in creep law – Eqns. (3.17) and (9.3))
n	(–)	work hardening exponent
P	(Pa)	pressure
Q	(J mole^{-1})	activation energy
R	(J K^{-1} mole^{-1})	universal gas constant
R	(m)	indenter radius
R	(m)	radius of curvature
R	(m)	surface roughness
r	(m)	radial distance
S	(Pa^{-1})	compliance (4th rank tensor)
S	(N m^{-1})	machine compliance (gradient of load–displacement curve – Eqn. (9.2))
S	(– or m^2)	sum of the squares of the residuals (for indenter depth – Eqns. (8.2) and (8.3))
S	(Pa)	stress amplitude during fatigue
s	(–)	slenderness ratio (Eqn. (10.8))
T	(–)	transform matrix
T	(K or °C)	temperature
T	(N m)	torque
t	(m)	wall thickness
t	(s)	time
U	(J m^3)	stored elastic strain energy
u	(–)	parameter relating to Miller indices for crystallographic directions
u	(m s^{-1})	velocity
v	(–)	parameter relating to Miller indices for crystallographic directions
W	(J)	energy of a dislocation (Eqn. (4.6))
W	(J)	energy released during crack advance (Eqn. (10.12))
w	(–)	parameter relating to Miller indices for crystallographic directions
w	(m)	width

Nomenclature

x	(m)	distance (Cartesian coordinate)
y	(m)	distance (Cartesian coordinate)
z	(m)	distance (Cartesian coordinate)
β	(m (Pa$\sqrt{m}$)$^{-n}$)	parameter in Paris–Erdogan fatigue law (Eqn. (10.20))
β	(° or radians)	angle subtended at the coil axis by a segment of coil (Fig. 10.3a)
Δ	(–)	relative change in volume (dilation – Eqn. (2.21))
δ	(m)	penetration, displacement or deflection
δ	(m)	crack opening displacement (Eqn. (10.19))
ε	(–)	strain (2nd rank tensor)
ε	(–)	dimensionless constant (Eqn. (9.2))
ϕ	(° or radians)	angle
γ	(–)	shear strain
γ	(J m^{-2})	surface energy
μ	(–)	coefficient of friction
κ	(m^{-1})	curvature
λ	(Pa)	variable in secular equation for stress (Eqn. (2.13))
λ	(° or radians)	angle (between slip direction and tensile axis)
Λ	(N or J m^{-1})	line tension or energy per unit length of a dislocation (Eqn. (4.7))
θ	(° or radians)	angle
ν	(–)	Poisson ratio
ρ	(m^{-2})	dislocation density
Σ	(N m^2)	beam stiffness
σ	(Pa)	stress (2nd rank tensor)
τ	(Pa)	shear stress (2nd rank tensor)
ξ	(–)	fraction of material transformed to martensite (Fig. 8.23)
ω	(–)	rotation (2nd rank tensor)
Ω	(–)	dimensionless length (Eqn. (10.10))

Subscripts

0	initial or reference
1	x direction
2	y direction
3	z direction
A	austenite
age	ageing
amb	ambient
B	Brinell
Braz	Brazier
c	critical

c	contact
cr	creep
f	finish
H	hydrostatic
i	indenter
i	suffix indicating direction
j	suffix indicating direction
K	Knoop
k	suffix indicating direction
L	Leeb
L	limit
M	martensite
max	maximum
N	nominal (engineering)
p	(equivalent) plastic
p	projected
p	polar
RA	Rockwell (category A)
RB	Rockwell (category B)
RC	Rockwell (category C)
ST	solution treatment
r	reduced
r	radial
s	start
T	true
V	Vickers
VC	Vickers cone
vM	von Mises
Y	yield
*	critical (e.g. fracture or ultimate tensile strength)

Acronyms

AFM	atomic force microscope
ASTM	American Society for Testing and Materials
bcc	body-centered cubic (crystal structure)
BSI	British Standards Institute
DIC	digital image correlation
DSC	differential scanning calorimetry
DoITPoMS	dissemination of IT for the promotion of materials science
EDM	electro-discharge machining
fcc	face-centered cubic (crystal structure)
FEM	finite element method
FIB	focussed ion beam (milling)
GP	Guinier–Preston (zones)

HCF	high cycle fatigue
hcp	hexagonal close-packed (crystal structure)
LCF	low cycle fatigue
LVDT	linear variable displacement transducer
MMC	metal matrix composites
OFHC	oxygen-free high conductivity (copper)
RA	reduction in area
SE	superelasticity
SEM	scanning electron microscope
SHPB	split Hopkinson pressure bar
SMA	shape memory alloys
SME	shape memory effect
TEM	transmission electron microscope
TSHB	torsional split Hopkinson bar
TLP	teaching and learning package
UTS	ultimate tensile strength (or stress)
UV	ultra-violet

1 General Introduction

This brief introductory chapter outlines the broad coverage of the book and its intended contribution in the context of other available sources. It is recognized that there are many excellent books covering the "mechanics of materials," often with a strong bias towards metals, but relatively few that are focussed strongly on testing procedures designed to reveal details about how they deform plastically. There are in fact many subtleties concerning metal plasticity and the information about it obtainable via various types of test. No attempt is made in this chapter to convey any of these, but the scene is set in terms of outlining the absolute basics of elastic and plastic deformation.

1.1 Rationale and Scope of the Book

The issue of how metals undergo plastic deformation (including creep – see the end of §1.3 below), and how this behavior can be measured and characterized, is a very old (and important!) one. Attempts to measure the "strength" of metals (and other materials) date back to antiquity. Understanding of the deformation mechanisms involved, and implications for systematic control over mechanical properties, are more recent, but they still extend back several decades. A number of texts were produced during the period in which most of the key advances were made, which was mainly between the 1950s and 1980s, and in many cases these have been followed up with a number of updated editions. Also, several new books have been published more recently. Many are strongly oriented towards engineering, with materials treated as (isotropic) continua and their properties described by analytical equations (constitutive laws). There are also several books containing a mixture of physical metallurgy and associated mechanical properties. Mechanical testing procedures are often covered in some way, although relatively few books have been dedicated to this area.

This book is partly aimed at comprehensive description of mechanical testing procedures, over a wide range of conditions (notably temperature and strain rate). Even the standard, "simple" procedures, such as uniaxial tensile testing, do in fact incorporate complexities that are not always well understood, but a large part of the motivation for the book relates to the fact that other testing geometries, particularly indentation, have become widespread recently. Also, the technique of finite element method (FEM) modeling, which is a powerful tool for investigation of mechanical deformation, is now mature and ubiquitous. Its use figures strongly in the book.

The theme of the book is not strongly oriented towards "nanoindentation" as such, although it is included in the coverage. It is mainly related to the obtaining of bulk mechanical properties (in the context of microstructure and deformation mechanisms, and their relationship to such properties). What would commonly be regarded as nanoindentation is not well suited to this objective, since the volume being mechanically interrogated is often too small for its response to be representative of the bulk. However, the key recent advances concerning indentation do not really relate to it being carried out on a fine scale, but rather to the process being instrumented so as to obtain detailed information about the (bulk) mechanical characteristics of the sample. This represents a fundamental advance, relative to its origins in hardness testing.

1.2 Structure and Readership of the Book

The structure of the book effectively comprises three sections. After this general introduction, there is a chapter dedicated to the handling of stresses and strains (as second rank tensors) and to the relationships between them during elastic deformation. This is followed by a chapter focussed on how such relationships are modified after the onset of plastic deformation (with the material treated as a homogeneous continuum) and then one on the mechanisms of plastic deformation and how they are affected by the microstructure of the metal. All three of these chapters can certainly be regarded as providing background information, although it is clearly relevant to a full understanding of plasticity. Similar material is available from a wide range of sources, but it is considered to be potentially helpful to include it in the book, for ease of reference. There is then a set of three chapters that cover the "standard" procedures of tensile, compressive and hardness testing. These are undoubtedly familiar to people involved with mechanical (plasticity) testing, but they do involve some subtleties that are not universally understood. These chapters contain guidelines for obtaining and interpreting experimental data most effectively with these procedures.

The final section of three chapters concerns testing techniques that have been developed more recently, with the first of these, outlining the current state of the art in indentation plastometry, given particularly close attention in view of its scope for obtaining detailed plasticity characteristics in a convenient and flexible way. The following one covers "nanoindentation," which has had a high profile over recent years, but actually suffers from some severe limitations in terms of obtaining information about bulk plasticity characteristics. The last chapter covers a range of fairly specialized test procedures, some in developmental form. There is thus a developing emphasis throughout the book on more recent (and research-oriented) topics, which is reflected in progressively increasing levels of citation of other published work. These references are collected at the end of each chapter. There is also a nomenclature listing at the beginning, with an attempt made to standardize the symbolism in use throughout the book.

It should be noted that the topic of fracture has a relatively low profile in the coverage. While indentation can sometimes be used to stimulate fracture, or at least crack propagation, it is not really an effect that can readily be investigated via indentation. Of course, some kind of fracture event does normally take place at the end of a tensile test, although local conditions during that event are often rather poorly defined. Testing aimed at obtaining fundamental fracture mechanics properties, such as the fracture energy or the fracture toughness, are carried out rather differently (and those tests are not described in the book). In fact, at least for most metals, it is largely the plastic deformation (yield stress and work hardening characteristics) that dictates their "strength" (as measured by the "ultimate tensile stress," UTS), with their fracture toughness as such (important as that is for other purposes) being of limited significance during such testing. Moreover, by interpreting the outcome of a tensile test via its simulation in an FEM model, it is possible to obtain an estimate of the critical strain for fracture, which is a widely used "property." Furthermore, coverage is included (in the final chapter) of certain types of cyclic loading test that may involve crack propagation and some of the basics of fracture mechanics are presented there. Such tests are closely related to real industrial service environments and, indeed, there is reference throughout the book to the relevance of various tests to industrial usage of metallic components.

A natural part of the readership is those already involved in mechanical testing (of metals), who need an update and a book for future reference. However, it is potentially of much broader appeal, since many people involved in the development of novel or improved materials and components have a need for understanding of the issues associated with mechanical deformation. Indentation plastometry has many attractions, and its usage is likely to increase over coming years. This book is aimed at providing a source of information for potential users of the technology. Furthermore, there is certainly potential for usage in university level teaching. The prerequisites are simply a very basic background in materials science and mechanics. Finally, there are many people involved in various manufacturing and processing sectors with an interest in improved understanding of the links between processing conditions and component performance in service. Metal plasticity and its characterization are central to this.

1.3 Basic Elastic and Plastic Property Ranges

There is, of course, a fundamental difference between elastic and plastic properties (of metals). Elastic deformation essentially arises because the distances between atoms are being changed (by an applied load). These distances increase in the direction of an applied tensile force, and decrease in directions transverse to this. Since the transverse reductions do not compensate fully for the axial extension, there is normally an associated volume change (unless the Poisson ratio has a value of 0.5, which is not the case for any metal). The "building blocks" (unit cells of the structure for crystalline materials, i.e. for the vast majority of metals) have all been distorted. Of course, the deformation is fully reversible on removing the applied load. This is the basic meaning of "elastic."

Plastic deformation, on the other hand, involves no volume change. It effectively occurs by moving some unit cells with respect to others (without distorting them). This most commonly occurs via the motion of dislocations (line defects within the structure). Other important differences, compared with the case of elastic deformation, are that the strains involved are commonly much greater and that, in general, the deformation is not readily reversible. (The word "plastic" effectively means "permanent.")

Details concerning plastic deformation – and there are many – are covered elsewhere in the book. However, it may be useful here to summarize the range likely to be encountered for the main parameter values (of metals). These are the Young's modulus (E), defined as the applied stress over the resultant (elastic) strain, the yield stress (applied stress level at which plastic deformation starts) and the ultimate tensile stress, which is the peak (nominal) stress, given by the maximum load attained during the test divided by the original sectional area of the sample. Typical values for various metals are shown in Table 1.1, although it should be emphasized that the yielding and strength values supplied there are simply indicative and are subject to large variations for the same material.

Table 1.1 Overview of the basic elastic and (approximate) plastic properties of a range of metals.

Material	Young's modulus, E (GPa)	Yield stress, σ_Y (MPa)	Yield strain, ε_Y (millistrain)	Ultimate tensile stress, σ_* (MPa)
Mild steel	205	250	1.2	370
304 stainless steel	200	250	1.2	600
4130 steel	205	400	1.9	700
Hadfield's Mn steel	210	350	1.7	1000
4340 steel	205	700	3.4	1100
A228 piano wire (spring steel)	200	1000	5	2000
Cast iron	170	350	2	500
1050 Al	71	100	1.4	150
6061 Al	69	250	3.6	300
7075 Al	71	500	7	600
OFHC Cu	130	350	2.7	370
Cu-2%Be	130	500	3.8	700
α/β brass (Cu-40%Zn)	97	250	2.5	400
Solder (Pb-40%Sn)	30	30	1	50
Inconel 718	200	600	3	1000
Ti-6Al-4V	115	950	8	1050
Be	300	250	0.8	370
Al-20%SiC$_p$ (MMC)	90	250	2.8	300

It may be noted that the Young's modulus (stiffness) depends only on the types of atoms present, which determines the interatomic forces. For example, all steels have a similar value of E, since they are all composed predominantly of Fe atoms. In contrast to this, the onset of plasticity (i.e. the yield stress) depends in a complex manner on the "microstructure" of the metal, a term that encompasses many features that can change with purity level, (relatively small) alloying additions, processing conditions, heat treatments etc. This arises mainly because these features are likely to have an effect on the ease of dislocation motion. Details of these effects are provided in Chapter 4. The yield stresses of different steels thus cover a wide range. This is also true of the ultimate tensile stress. (In fact, for much of the data in Table 1.1, a strong caveat should be appended to the yield stress and tensile strength values, to the effect that they could change significantly if the component concerned were to be heated or subjected to some other kind of treatment.) It also follows that the strain at the onset of yielding varies between different metals, although, as can be seen, the value is in all cases quite small ($<\sim 1\%$, i.e. 10 millistrain). In contrast, plastic strains are commonly large (typically several tens of %).

At this juncture, only a few broad points of this type should be noted concerning the data in Table 1.1. Reliable and meaningful information concerning the plasticity of metals is far from simple in terms of both experimental procedures and interpretation of the data. Of course, the values in the table do refer to material response at ambient temperature. It will tend to be different at both lower and (particularly) higher temperatures than this. It should also be noted that the behavior is assumed to be time-independent, so the rate at which the load is applied is taken to have no effect. In practice, this is often fairly reliable over the range of loading rates that is likely to be used, but both very high rates and very low rates can lead to a significantly different response. In particular, what is usually termed creep – i.e. progressive plastic deformation at a constant applied load – may become noticeable at low loading rates. However, such behavior should be characterized using laws that incorporate time, rather than making some sort of adjustment to the apparent plasticity parameters. All of these issues are covered in detail elsewhere in the book.

A final point can be made that concerns composite materials. Of course, most composites are based on polymers, which are outside the scope of the book, but **metal matrix composites** (MMC) are in industrial use and are certainly worthy of study. In most cases, they contain ceramic reinforcement, in the form of either particles or fibers. These raise the stiffness and also usually lead to a higher yield stress and UTS, although often at a cost in reduced ductility and toughness. Table 1.1 does contain data for an MMC, although such materials have not really reached the levels of development and stability typical of many alloys, so this material, and these values, are not very well defined. It is perhaps also worth recognizing that metals containing relatively high levels of *porosity* can be treated as a special type of composite. Metal components, particularly castings, do sometimes contain porosity, although in many cases the levels are too low ($<\sim 1\%$) to have much effect on elastic or plastic properties. On the other hand, metals containing high levels of porosity ($>\sim 30\%$), produced deliberately in some way, have found some applications: they would commonly be described as metallic foams. Their properties, both elastic and plastic, are naturally very different from the corresponding pore-free metal. Again, however, such materials are not well defined or mature, so it's difficult to justify the inclusion of a specific example in Table 1.1. Nevertheless, it is worth noting that the presence of pores in a metal, at or above a level of a few %, can affect the mechanical response.

2 Stresses, Strains and Elasticity

Comprehensive treatment of metal plasticity requires an understanding of the fundamental nature of stresses and strains. A stress can be understood at a basic level as a force per unit area on which it acts, while a strain is an extension divided by an original length. However, the limitations of these definitions rapidly become clear when considering anything other than very simple loading situations. Analysis of various practical situations can in fact be rigorously implemented without becoming embroiled in mathematical complexity, most commonly via usage of commercial (finite element) numerical modeling packages. However, there are various issues involved in such treatments, which need to be appreciated by practitioners if outcomes are to be understood in detail. This chapter covers the necessary fundamentals, relating to stresses and strains, and to their relationship during elastic (reversible) deformation. How this relationship becomes modified when the material undergoes plastic (permanent) deformation is covered in the following chapter.

2.1 Stress and Strain as Second Rank Tensors

Although stress and strain can sometimes be handled as if they were simple scalars (i.e. numbers with no directions associated with them), more rigorous treatment is often required. Stress and strain are in fact second rank tensors. The utility of tensors is mainly concerned with treating differences in the response or characteristics of a material in different directions within it. Their usage is thus particularly required when treating **anisotropic** materials. However, even for isotropic materials, tensor analysis is necessary, or at least very helpful, when treating many types of mechanical loading.

A tensor is an n-dimensional array of values, where n is the "rank" of the tensor. The simplest type is thus a tensor of zeroth rank, which is a *scalar* – i.e. a single numerical value. Properties like temperature and density are scalars. They are not associated with any particular direction in the material concerned, and the variable does not require any associated index (suffix[1]). A first rank tensor is a *vector*. This is a 1-D array of values. There are normally three values in the array, each corresponding to one of three (orthogonal) directions. Each value has a single suffix, specifying the

---

[1] The term "suffix" is in common use for these indices, although they are employed as subscripts.

direction concerned. These suffices are commonly numerical (1, 2 and 3), although sometimes other nomenclature, such as (x, y and z) or (r, θ and z), may be used. Force and velocity are examples of vectors. The components of a vector can thus be written down in a form such as

$$\boldsymbol{F} = F_i = [F_1 \ F_2 \ F_3] \tag{2.1}$$

with each of the suffices (1, 2 and 3) referring to a specific direction, such as (x, y and z).

There are other variables, including **stress**, for which each component requires the specification of two directions, rather than one, so that two suffices are needed. In the case of stress, these two suffices specify, firstly, the direction in which a force is being applied and, secondly, the normal of the plane on which the force is acting. Stress is thus a **second rank tensor** and the components form a 2-D array of values.

$$\sigma_{ij} = \begin{bmatrix} \sigma_{11} & \sigma_{12} & \sigma_{13} \\ \sigma_{21} & \sigma_{22} & \sigma_{23} \\ \sigma_{31} & \sigma_{32} & \sigma_{33} \end{bmatrix} \tag{2.2}$$

When the suffices i and j are the same, the force acts parallel to the plane normal, and so the component concerned is a **normal stress**. When they are different, it is a **shear stress** (and sometimes the symbol τ is used instead of σ for such components).

Some of the stresses that could act on a body are depicted in Fig. 2.1. Provided that the body is in **static equilibrium**, which is commonly assumed, then the normal forces acting on opposite faces so as to generate a normal stress (e.g. σ_{33} in Fig. 2.1) must be equal in magnitude and anti-parallel in direction. (If this were not the case, then the body would **translate**.) For shear stresses, a further condition applies. Not only must the two forces generating the σ_{23} stress (see Fig. 2.1) be equal and opposite, but the magnitude of the σ_{32} stress must be equal to that of the σ_{23} stress. (If this were not the

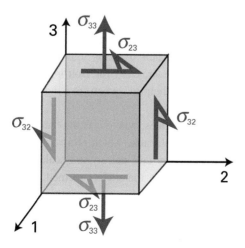

Fig. 2.1 Illustration of the nomenclature of stresses acting on a body.

case, then the body would *rotate*.) Shear stresses thus act in pairs. This applies to all shear stresses, so that $\sigma_{ij} = \sigma_{ji}$, and the tensor represented in Eqn. (2.2) must be *symmetrical*.

2.2 Transformation of Axes

It follows that there are just six independent components in a general stress state – three normal stresses and three shear stresses. Their magnitude will, of course, depend on the directions of the axes chosen to provide the frame of reference. However, the state of stress itself will clearly be unaffected if we choose an alternative frame of reference. Any tensor can be *transformed* so as to be referred to a new set of axes, provided the orientation of these with respect to the original set is specified. Furthermore, any stress state can be expressed solely in terms of normal stresses (i.e. all shear stresses are zero), provided that a certain set of axes is chosen. Partly because it's often helpful to express a stress state in terms of this unique set of normal stresses, the procedures for transforming tensors are important. They are illustrated first for a vector (force) and then for a stress.

2.2.1 Transforming First Rank Tensors (Vectors)

Consider a vector (a force, for example), $F (= [0, F_2, F_3])$, with components that are referred to the axis set (1, 2, 3). A specific reorientation of this set of axes is now introduced, namely a rotation by an angle ϕ about the 1-axis, to create a new axis set $(1', 2', 3')$ – see Fig. 2.2. In this case, the new $1'$-axis coincides with the old 1-axis, but the $2'$- and $3'$-axes have been rotated with respect to the 2- and 3-axes.

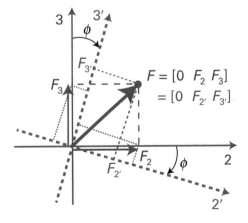

Fig. 2.2 Rotation, in the 2–3 plane, of the axes forming the reference frame for a vector F.

The values of $F_{2'}$ and $F_{3'}$ are found by resolving the components F_2 and F_3 onto the 2' and 3' axes and adding these resolved components together:

$$F_{2'} = F_2 \cos(2'-2) + F_3 \cos(2'-3)$$
$$F_{3'} = F_2 \cos(3'-2) + F_3 \cos(3'-3) \quad (2.3)$$

where the symbolism $(x-y)$ represents the angle between the x and y axes. In terms of the angle ϕ, these two equations can be written as

$$F_{2'} = F_2 \cos(-\phi) + F_3 \cos(-90-\phi) = F_2 \cos\phi - F_3 \sin\phi$$
$$F_{3'} = F_2 \cos(90-\phi) + F_3 \cos(-\phi) = F_2 \sin\phi + F_3 \cos\phi \quad (2.4)$$

Clearly, these cosines (of angles between new and old axes) are central to such transformations. They are commonly termed **direction cosines** and represented by a_{ij}, which is conventionally the cosine of the angle between the new i direction ($=i'$) and the old j direction. Of course, the rationale can be extended to cases in which all three axes have been reoriented, leading to the following set of equations:

$$F_{1'} = a_{11}F_1 + a_{12}F_2 + a_{13}F_3$$
$$F_{2'} = a_{21}F_1 + a_{22}F_2 + a_{23}F_3 \quad (2.5)$$
$$F_{3'} = a_{31}F_1 + a_{32}F_2 + a_{33}F_3$$

It can be seen that the direction cosines form a matrix and this set of equations can be written more compactly in matrix form:

$$\begin{bmatrix} F_{1'} \\ F_{2'} \\ F_{3'} \end{bmatrix} = [T] \begin{bmatrix} F_1 \\ F_2 \\ F_3 \end{bmatrix} \quad (2.6)$$

in which the **transformation matrix** is given by

$$[T] = \begin{bmatrix} a_{11} & a_{12} & a_{13} \\ a_{21} & a_{22} & a_{23} \\ a_{31} & a_{32} & a_{33} \end{bmatrix} \quad (2.7)$$

Sets of equations such as Eq. (2.6) can be written even more concisely by using the **Einstein summation convention**. This states that, when a suffix occurs twice in the same term, then this indicates that summation should be carried out with respect to that term. For example, in the equation

$$F_{i'} = a_{ij}F_j \quad (2.8)$$

j is a **dummy suffix**, which is to be summed (from 1 to 3). The i suffix, on the other hand, is a **free suffix**, which can be given any chosen value. For example, Eqn. (2.8) could be used to create the equation

$$F_{1'} = a_{11}F_1 + a_{12}F_2 + a_{13}F_3 \quad (2.9)$$

and also the two other equations, corresponding to i being equal to 2 or 3.

2.2.2 Transforming Second Rank Tensors and Principal Stresses

Extension of the above treatment to second rank tensors, such as stress, follows quite logically. However, for a stress, we are concerned not just with resolving three single components into new directions (e.g. F_1, F_2 and F_3 in the case of a force), but rather we need to take account of the fact that **both the force** and **the area on which it is acting** will **change** when we refer them to different axes. In other words, the operation we carried out with respect to a single suffix in the case of force, F_i, has to be implemented with respect to two suffices for a stress, σ_{ij}. Each stress component thus needs to be multiplied by **two direction cosines**, rather than one. The set of equations corresponding to Eqn. (2.8) can thus be written

$$\sigma'_{ij} = a_{ik} a_{jl} \sigma_{kl} \qquad (2.10)$$

in which the prime notation is now applied to the symbol, rather than the individual suffices, to denote the transformed version. Equation (2.10) can be expanded into nine equations, each giving one of the nine components of the stress tensor when referred to the new set of axes (although only six are required, since the stress tensor is always symmetrical). Both k and l are dummy suffices in this equation (since they are repeated), while i and j are free suffices. The first equation of the set, corresponding to $i = 1$ and $j = 1$, can thus be expanded to

$$\sigma'_{11} = \begin{array}{l} a_{11}a_{11}\sigma_{11} + a_{11}a_{12}\sigma_{12} + a_{11}a_{13}\sigma_{13} \\ a_{12}a_{11}\sigma_{21} + a_{12}a_{12}\sigma_{22} + a_{12}a_{13}\sigma_{23} \\ a_{13}a_{11}\sigma_{31} + a_{13}a_{12}\sigma_{32} + a_{13}a_{13}\sigma_{33} \end{array}$$

while that corresponding to $i = 3$ and $j = 2$ is given by

$$\sigma'_{32} = \begin{array}{l} a_{31}a_{21}\sigma_{11} + a_{31}a_{22}\sigma_{12} + a_{31}a_{23}\sigma_{13} \\ a_{32}a_{21}\sigma_{21} + a_{32}a_{22}\sigma_{22} + a_{32}a_{23}\sigma_{23} \\ a_{33}a_{21}\sigma_{31} + a_{33}a_{22}\sigma_{32} + a_{33}a_{23}\sigma_{33} \end{array}$$

One motivation for treating stresses within this mathematical framework is that it facilitates identification of the **principal stresses**. These are the normal stresses acting on the **principal planes**, which are the planes on which there are no shear stresses. (These principal stresses are the **eigenvalues** of the stress tensor.) Any state of stress can thus be transformed such that it can be expressed in the form

$$\sigma_{ij} = \begin{bmatrix} \sigma_1 & 0 & 0 \\ 0 & \sigma_2 & 0 \\ 0 & 0 & \sigma_3 \end{bmatrix} \qquad (2.11)$$

with the single suffix being commonly used to denote a principal stress. (It's important to recognize, however, that these are still second rank tensors, and should thus, strictly speaking, always have two suffices.) Obtaining these principle stresses, for a general 3-D stress state, requires **diagonalizing** of the stress tensor. This is done by setting the **determinant** of the coefficients equal to zero. The solutions for the principal stresses are thus found from

$$\begin{bmatrix} \sigma_{11} - \lambda & \sigma_{12} & \sigma_{13} \\ \sigma_{21} & \sigma_{22} - \lambda & \sigma_{23} \\ \sigma_{31} & \sigma_{32} & \sigma_{33} - \lambda \end{bmatrix} = 0 \qquad (2.12)$$

This is a cubic equation in λ, the roots of which are the principal stresses. It is often termed the *secular equation*. Expanding the terms in Eqn. (2.12) leads to

$$\lambda^3 - I_1\lambda^2 + I_2\lambda - I_3 = 0 \qquad (2.13)$$

in which the coefficients (termed the *invariants*, since they do not vary as the axes are changed) are given by

$$\begin{aligned} I_1 &= \sigma_{11} + \sigma_{22} + \sigma_{33} \\ I_2 &= \sigma_{11}\sigma_{22} + \sigma_{22}\sigma_{33} + \sigma_{33}\sigma_{11} - \sigma_{12}^2 - \sigma_{23}^2 - \sigma_{31}^2 \\ I_3 &= \sigma_{11}\sigma_{22}\sigma_{33} + 2\sigma_{12}\sigma_{23}\sigma_{31} - \sigma_{11}\sigma_{23}^2 - \sigma_{22}\sigma_{13}^2 - \sigma_{33}\sigma_{12}^2 \end{aligned} \qquad (2.14)$$

2.2.3 Use of Mohr's Circle

It's therefore fairly straightforward, if a little cumbersome, to find the principal stresses, and the orientation of the (normals of the) principal planes, for a 3-D stress state specified with respect to an arbitrary set of axes. It just requires a cubic equation to be solved. However, in practice it is common to know one principle plane (direction), but to be interested in finding the principal directions within that plane, or in establishing how the normal and shear stresses vary with direction in that plane. Provided the plane concerned is a principal one, this problem reduces to solving a quadratic equation, rather than a cubic. It can therefore be tackled via a geometrical construction, equivalent to using some simple trigonometry. This is the basis of Mohr's circle, which was proposed in 1892 by Christian Otto Mohr, a German civil engineer who became a professor of mechanics in Dresden.

Provided the 1–2 plane is a principal plane, the appropriate version of Eqn. (2.12) is

$$\begin{bmatrix} \sigma_{11} - \lambda & \sigma_{12} & 0 \\ \sigma_{21} & \sigma_{22} - \lambda & 0 \\ 0 & 0 & \sigma_{33} - \lambda \end{bmatrix} = 0$$

and one principal stress is clearly σ_{33} ($= \sigma_3$). Since $\sigma_{12} = \sigma_{21}$, the secular equation reduces to

$$(\sigma_{11} - \lambda)(\sigma_{22} - \lambda) - \sigma_{12}^2 = 0$$
$$\lambda^2 - (\sigma_{11} + \sigma_{22})\lambda + (\sigma_{11}\sigma_{22} - \sigma_{12}^2) = 0$$

This has the solution

$$\begin{aligned} \sigma_1 &= \left(\frac{\sigma_{11} + \sigma_{22}}{2}\right) + \sqrt{\left(\frac{\sigma_{11} - \sigma_{22}}{2}\right)^2 + \sigma_{12}^2} \\ \sigma_2 &= \left(\frac{\sigma_{11} + \sigma_{22}}{2}\right) - \sqrt{\left(\frac{\sigma_{11} - \sigma_{22}}{2}\right)^2 + \sigma_{12}^2} \end{aligned} \qquad (2.15)$$

It's also possible to start with the tensor in diagonalized form and then transform it to give σ'_{11}, σ'_{22} and σ'_{12}, for a specified orientation

$$\begin{bmatrix} \sigma'_{11} & \sigma'_{12} & 0 \\ \sigma'_{12} & \sigma'_{22} & 0 \\ 0 & 0 & \sigma_3 \end{bmatrix} = [T] \begin{bmatrix} \sigma_1 & 0 & 0 \\ 0 & \sigma_2 & 0 \\ 0 & 0 & \sigma_3 \end{bmatrix}$$

Since we are rotating about a principal axis, the form of Eqn. (2.10) applicable in this case reduces to

$$\sigma'_{11} = a_{11}a_{11}\sigma_1 + a_{12}a_{12}\sigma_2 + a_{13}a_{13}\sigma_3$$
$$\sigma'_{22} = a_{21}a_{21}\sigma_1 + a_{22}a_{22}\sigma_2 + a_{23}a_{23}\sigma_3$$
$$\sigma'_{12} = a_{11}a_{21}\sigma_1 + a_{12}a_{22}\sigma_2 + a_{13}a_{23}\sigma_3$$

The set of direction cosines applicable here – see Eqn. (2.4) – can be written as

$$\begin{bmatrix} a_{11} & a_{12} & a_{13} \\ a_{21} & a_{22} & a_{23} \\ a_{31} & a_{32} & a_{33} \end{bmatrix} = \begin{bmatrix} \cos\phi & -\sin\phi & 0 \\ \sin\phi & \cos\phi & 0 \\ 0 & 0 & 1 \end{bmatrix}$$

so it follows that these stresses are given by

$$\sigma'_{11} = \cos^2\phi\,\sigma_1 + \sin^2\phi\,\sigma_2$$
$$\sigma'_{22} = \sin^2\phi\,\sigma_1 + \cos^2\phi\,\sigma_2$$
$$\sigma'_{12} = \sin\phi\cos\phi\,\sigma_1 - \sin\phi\cos\phi\,\sigma_2$$

These equations can also be written in a form involving 2ϕ, rather than ϕ:

$$\sigma'_{11} = \left(\frac{\sigma_1 + \sigma_2}{2}\right) + \left(\frac{\sigma_1 - \sigma_2}{2}\right)\cos 2\phi$$
$$\sigma'_{22} = \left(\frac{\sigma_1 + \sigma_2}{2}\right) - \left(\frac{\sigma_1 - \sigma_2}{2}\right)\cos 2\phi \qquad (2.16)$$
$$\sigma'_{12} = \left(\frac{\sigma_1 - \sigma_2}{2}\right)\sin 2\phi$$

It can now be seen how these equations can be solved using a geometrical construction. These expressions (i.e. the normal and shear stresses on a plane rotated by an angle ϕ from that on which σ_1 acts) are given by the coordinates of a point rotated by 2ϕ around the circumference of a circle centered at the mean of the two principal stresses, and with a radius equal to half their difference. This is illustrated in Fig. 2.3. This construction provides a convenient method of calculating the stresses acting on particular planes, establishing principal stresses and their orientations, finding the planes on which peak shear stresses[2] operate etc.

---

[2] Confusion occasionally arises about the sign of shear stresses (and strains), and it is sometimes implied that the Mohr circle has positive and negative sides in the vertical direction. However, there is no need for this, since the sign of a shear stress (or strain) is solely an issue of convention, and has no physical significance (unlike the sign of a normal stress or strain). Since both stress and strain tensors are symmetric, it is simplest to take all shear ($i \neq j$) terms as always positive. The only possible issue is that a consistent convention may be needed regarding the sense of rotation in Mohr space and physical space.

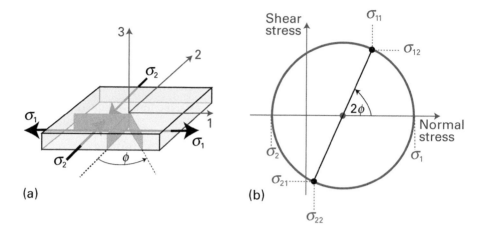

Fig. 2.3 Mohr's circle construction for a principal plane (1–2 plane), showing (a) principal stresses in the plane (σ_1 and σ_2, for a case in which σ_1 is tensile and σ_2 compressive) and (b) corresponding Mohr's circle, giving, in addition to the principal stresses, the normal and shear stresses acting on a plane rotated about the 3-axis by an angle ϕ from that on which σ_1 acts.

2.3 Representation of Strain

The application of a set of stresses causes all points in the body to be displaced, relative to their initial locations, so that their coordinates, in a given frame of reference, will change. Provided that the body is *macroscopically homogeneous*, these displacements can be related to the stress state, via a description of the elastic properties of the material, which may, of course, be anisotropic. The **relative displacement tensor** (sometimes, rather confusingly, called the **deformation tensor**) indicates how any point in the body becomes displaced. It is a second rank tensor, with the first suffix representing the direction in which the displacement has occurred and the second one the reference direction. This is illustrated in Fig. 2.4. All of the terms are (dimensionless) ratios of two distances. The shear terms (e_{ij}, with $i \neq j$) can also be considered as angles – since they are small, they are approximately equal to their tangents. Care is needed in identifying the **strain tensor**, since an applied set of stresses will in general create some **rigid body rotation**, as well as genuine deformation (shape change) of the body. The strain tensor reflects only the shape change.

The sign of a shear component of the relative displacement tensor is taken as positive when the positive axis is rotated towards the positive direction of the other axis. For example, in Fig. 2.4(b) the e_{23} component would be positive. Separation of the strain from the rigid body rotation is illustrated in Fig. 2.5, for shear deformation in the y–z (2–3) plane.

This separation can be expressed more generally. Any second rank tensor can be separated into symmetrical and anti-symmetrical components. For the relative displacement tensor, this can be written as

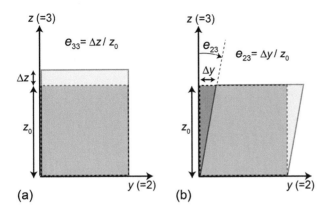

Fig. 2.4 Illustration, in the y–z (2–3) plane, of how the terms of the relative displacement tensor are defined, showing (a) a normal term, e_{33}, and (b) a shear term, e_{23}.

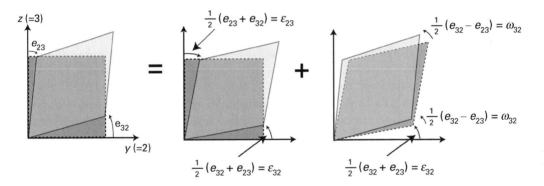

Fig. 2.5 Illustration of how, in the 2–3 plane, the relative displacements e_{23} and e_{32} can be represented as the sum of a strain, ε_{23} (= ε_{32}) and a rigid body rotation, ω_{23} (= $-\omega_{32}$).

$$e_{ij} = \varepsilon_{ij} + \omega_{ij} = \frac{1}{2}\left(e_{ij} + e_{ji}\right) + \frac{1}{2}\left(e_{ij} - e_{ji}\right) \quad (2.17)$$

Since the **strain**, ε_{ij}, is a symmetrical tensor, then

$$\varepsilon_{ij} = \frac{1}{2}\left(e_{ij} + e_{ji}\right) = \varepsilon_{ji} \quad (2.18)$$

whereas for ω_{ij}, the **rotation tensor**, which is anti-symmetrical,

$$\omega_{ij} = \frac{1}{2}\left(e_{ij} - e_{ji}\right) = \omega_{ji} \quad (2.19)$$

It's readily shown that such an anti-symmetrical tensor represents only a rotation. For example, in the 2–3 plane, the components of such a tensor can be written

$$\omega_{ij} = \begin{bmatrix} 0 & e_{23} \\ -e_{23} & 0 \end{bmatrix} \quad (2.20)$$

It can be seen in Fig. 2.6 that this represents solely rigid body rotation.

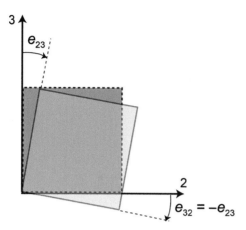

Fig. 2.6 Illustration of how an anti-symmetrical relative displacement tensor, in the 2–3 plane, represents only a rotation, with the body not being subject to any strain.

As in the case of a stress tensor, a strain tensor can be diagonalized to give the ***principal strains***. These are the normal strains in the principal directions. As with stress, the principal directions are a unique set of (orthogonal) plane normals. For these directions, there are no shear strains. A strain tensor can be manipulated in a very similar way to a stress tensor. For example, Mohr's circle can be used to find the principal strains or to facilitate other calculations or visualizations concerning the strain.

A point worth noting at this stage is that the strain tensor can itself be divided into two components – one representing the change in volume of the body and the other the change in its shape. These are commonly termed the **hydrostatic** and **deviatoric** components. Consider a cube subjected to principal strains ε_1, ε_2 and ε_3. The change in volume, often called the **dilation**, can be written as

$$\Delta = (1+\varepsilon_1)(1+\varepsilon_2)(1+\varepsilon_3) - 1 \sim \varepsilon_1 + \varepsilon_2 + \varepsilon_3 \tag{2.21}$$

In fact, we can write $\Delta = \varepsilon_{ii}$ for any strain tensor, since the sum of the diagonal terms of a second rank tensor is an invariant, and hence independent of the reference axes – see Eqn. (2.14). The hydrostatic strain tensor has one third of the dilation for all three of the normal terms, and no shear terms. The residual part is the deviatoric component:

$$\begin{bmatrix} \varepsilon_{11} & \varepsilon_{12} & \varepsilon_{13} \\ \varepsilon_{21} & \varepsilon_{22} & \varepsilon_{23} \\ \varepsilon_{31} & \varepsilon_{32} & \varepsilon_{33} \end{bmatrix} = \begin{bmatrix} \Delta/3 & 0 & 0 \\ 0 & \Delta/3 & 0 \\ 0 & 0 & \Delta/3 \end{bmatrix} + \begin{bmatrix} \varepsilon_{11} - \Delta/3 & \varepsilon_{12} & \varepsilon_{13} \\ \varepsilon_{21} & \varepsilon_{22} - \Delta/3 & \varepsilon_{23} \\ \varepsilon_{31} & \varepsilon_{32} & \varepsilon_{33} - \Delta/3 \end{bmatrix} \tag{2.22}$$

2.4 Stress–Strain Relationships and Engineering Constants

2.4.1 Stiffness and Compliance Tensors

The tensors we have been treating so far have all been ***field tensors***. These are imposed on bodies, or regions of space, in some way. Relationships between field tensors, for example between stress and resultant strain, depend on properties on the material concerned. These are characterized by **matter tensors**. Matter tensors reflect the **symmetry** exhibited by the material. The rank of a matter tensor is equal to the sum of the ranks of the two field tensors that it links.

The relationship between a stress (tensor) and the strain (tensor) generated by it can be written

$$\sigma_{ij} = C_{ijkl}\varepsilon_{kl} \tag{2.23}$$

where C_{ijkl} is the **stiffness tensor**. This is a fourth rank tensor, with 3^4 ($= 81$) components. (The rank of a tensor is equal to the number of its suffices, although there are some situations in which a convention may be used such that this is not the case – an example is provided by principal stresses and strains being given a single suffix.) Equation (2.23) is a generalized expression of **Hooke's law**. It represents nine equations, generated according to the Einstein summation convention. For example, the first of these is

$$\sigma_{11} = \begin{matrix} C_{1111}\varepsilon_{11} + C_{1112}\varepsilon_{12} + C_{1113}\varepsilon_{13} \\ C_{1121}\varepsilon_{21} + C_{1122}\varepsilon_{22} + C_{1123}\varepsilon_{23} \\ C_{1131}\varepsilon_{31} + C_{1132}\varepsilon_{32} + C_{1133}\varepsilon_{33} \end{matrix} \tag{2.24}$$

The stress–strain relationship can also be expressed in the inverse sense as

$$\varepsilon_{ij} = S_{ijkl}\sigma_{kl} \tag{2.25}$$

in which S_{ijkl} is the **compliance tensor**. (The symbols conventionally used for stiffness and compliance are the reverse of the initial letters of these words.)

2.4.2 Relationship to Elastic Constants

The above expressions look a little cumbersome and daunting, but in practice the treatment can be simplified. The symmetry of stress and strain tensors when the body is in static equilibrium means that

$$C_{ijkl} = C_{ijlk} = C_{jikl} = C_{jilk} \tag{2.26}$$

reducing the number of independent components from 81 to 36. Furthermore, the symmetry exhibited by the material commonly results in further reductions in this number. In fact, for an isotropic material, it is reduced to just 2, so the elastic behavior of isotropic materials is fully specified by the values of two constants.

In terms of engineering constants, the relationship between (normal) stress and strain can be obtained by considering the application of a single stress σ_1, generating a strain ε_1, so that

$$\sigma_1 = E\varepsilon_1 \tag{2.27}$$

in which E is the **Young's modulus**. One might be tempted to deduce that E is given by C_{1111}. However, this is incorrect, since the application of a stress σ_1 generates, not only a (direct) strain ε_1, but also **Poisson strains** ε_2 and ε_3 (see below), which would appear in the full (tensorial) equation for σ_1. Expressed in terms of compliance, however, the equation we need (from Eqn. (2.25)) is

$$\varepsilon_1 = S_{1111}\sigma_1$$

$$E = \frac{1}{S_{1111}} \left(= \frac{1}{S_{2222}} = \frac{1}{S_{3333}} \right) \tag{2.28}$$

The second elastic constant specified for an isotropic material is commonly the **Poisson ratio**, ν. This gives the transverse contraction strain that accompanies an axial extension strain. Again considering a single (normal) stress σ_1, generating principal strains ε_1, ε_2 and ε_3, the Poisson ratio is defined by

$$\nu = \frac{-\varepsilon_2}{\varepsilon_1} = \frac{-\varepsilon_3}{\varepsilon_1} \tag{2.29}$$

The tensorial expression for ε_2, with only σ_1 applied, is

$$\varepsilon_2 = S_{2211}\sigma_1 (= S_{1122}\sigma_1)$$

so that

$$\nu = \frac{-\varepsilon_2}{\varepsilon_1} = \frac{-S_{1122}\sigma_1}{(\sigma_1/E)} = -ES_{1122} \tag{2.30}$$

Since Poisson strains simply superimpose during multi-axial loading, we can write the following expressions for the (principal) strains arising from a set of (principal) stresses:

$$\begin{aligned}\varepsilon_1 &= \left(\frac{\sigma_1 - \nu(\sigma_2 + \sigma_3)}{E}\right) \\ \varepsilon_2 &= \left(\frac{\sigma_2 - \nu(\sigma_1 + \sigma_3)}{E}\right) \\ \varepsilon_3 &= \left(\frac{\sigma_3 - \nu(\sigma_1 + \sigma_2)}{E}\right)\end{aligned} \tag{2.31}$$

2.4.3 Engineering Shear Strains, Shear Modulus and Bulk Modulus

A minor complication arises when considering shear strains and the shear modulus. One might imagine that the **shear modulus**, G, would be defined as shear stress divided by shear strain, e.g. $\sigma_{12}/\varepsilon_{12}$. Unfortunately, $G \neq \sigma_{12}/\varepsilon_{12}$. The problem is associated with the difference between a **tensorial shear strain**, such as ε_{12}, and the corresponding **engineering shear strain**, γ_{12}. The shear modulus is conventionally defined by

$$G = \frac{\sigma_{12}}{\gamma_{12}} \tag{2.32}$$

in which the value of γ_{12} is simply the measured strain, with no account being taken of the fact that it really represents the sum of two shear strains, one being

2.4 Stress–Strain Relations and Engineering Constants

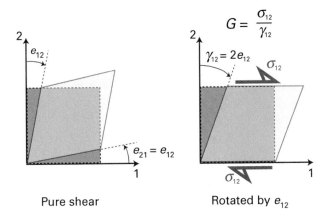

Fig. 2.7 Illustration of how a pure shear deformation, in the 1–2 plane, when rotated about the 3-axis, can be viewed as a simple shear, and used to define the shear modulus, G, in terms of the engineering shear strain, γ_{12}. The presence of the other shear stress acting in this plane, σ_{21}, is neglected in this definition.

generated by each of the pair of shear stresses which is being applied. This is illustrated in Fig. 2.7.

The array of engineering strains (ε_{ii} and γ_{ij}) thus do not constitute a genuine tensor and cannot be transformed using the standard procedures for second rank tensors. However, we can still define the shear modulus in terms of components of the compliance tensor. With only σ_{12} and σ_{21} acting, Eqn. (2.25) gives

$$\varepsilon_{12} = S_{1212}\sigma_{12} + S_{1221}\sigma_{21} = 2S_{1212}\sigma_{12}$$
$$G = \frac{\sigma_{12}}{\gamma_{12}} = \frac{2\sigma_{12}}{\varepsilon_{12}} = \frac{1}{S_{1212}} \quad (2.33)$$

In fact, similar procedures can be used to establish all of the components of the compliance and stiffness tensors, in terms of (measured) engineering elastic constants. Of course, the procedure is more complex for anisotropic materials, for which there are more than two independent elastic constants, but it can still be carried out – see below.

It may finally be noted that the bulk modulus, K, is defined by

$$K = \frac{\sigma_H}{\Delta} = \frac{\frac{1}{3}(\sigma_1 + \sigma_2 + \sigma_3)}{(\varepsilon_1 + \varepsilon_2 + \varepsilon_3)} \quad (2.34)$$

where σ_H is the hydrostatic component of the stress state. The bulk modulus is a measure of the resistance of the material to volume change

2.4.4 Relationships between Elastic Constants

Since only two elastic constants are required to fully define the behavior of an isotropic material, it follows that there must be inter-relationships between the four that have been considered here. For example, adding up the three equations in Eqn. (2.31) gives

$$\varepsilon_1 + \varepsilon_2 + \varepsilon_3 = \left(\frac{1-2\nu}{E}\right)(\sigma_1 + \sigma_2 + \sigma_3)$$

$$\Delta = \left(\frac{1-2\nu}{E}\right)3\sigma_H \qquad (2.35)$$

$$K = \frac{\sigma_H}{\Delta} = \frac{E}{3(1-2\nu)}$$

Other relationships are commonly quoted; for example, the shear modulus is given by

$$G = \frac{E}{2(1+\nu)} \qquad (2.36)$$

3 Continuum Plasticity

The handling of stress and strain during elastic deformation is covered in the preceding chapter. However, the situation becomes more complex after the onset of plastic deformation. Whereas elastic straining essentially occurs just via changes in interatomic spacing, the mechanisms involved in plastic (permanent) deformation are far from simple. These mechanisms are described in some detail in the next chapter. The current chapter is based, as is the previous one, on treating the material as a homogeneous continuum, albeit one that may be anisotropic (i.e. exhibit different responses in different directions). Much of the coverage concerns conditions for the onset of plasticity (often described as "yielding") and subsequent rises in applied stress that are required for further plastic straining ("work hardening"). Two yielding criteria are in common use and these are described. The work hardening behavior is often quantified using empirical constitutive laws and two of the most prominent of these are also outlined. This chapter also covers the representation of temporal effects – both the changes in stress–strain characteristics that occur when high strain rates are imposed and the progressive straining that can take place over long periods under constant stress, which is often termed "creep."

3.1 The Onset of Plasticity and Yielding Criteria

Chapter 2 relates to elastic (reversible) deformation. Of course, the elastic behavior of metals is of considerable importance, but their plastic deformation characteristics are in many ways even more important, and they certainly introduce higher levels of complexity – both in terms of the mechanisms responsible and from the point of view of their analytical representation. In summary, plastic deformation of metals most commonly occurs as a result of the ***glide of dislocations***, driven by shear stresses. (In some cases, ***deformation twinning*** may contribute, but this also requires shear stresses in a similar way, and also involves no volume change.) These, and other, mechanisms of plasticity are covered in Chapter 4. In a ***polycrystal*** (i.e. in most metallic samples), individual grains must deform in a cooperative way, so that each undergoes a relatively complex shape change (requiring the operation of ***multiple slip systems***), consistent with those of its neighbors.

In general, the continuation of plastic deformation requires a progressively increasing level of applied stress. This effect is termed "***work hardening***" or "***strain hardening***." Broadly, it arises because, as more dislocations are created, and as they interact with each other (creating jogs and tangles), they tend to become less mobile. Again, these issues are covered in mechanistic terms in Chapter 4, and are treated here simply as effects that can be represented via equations (constitutive laws). In the rest of the current chapter, issues related to microstructure and the micro-mechanisms of plasticity are ignored and the material is simply treated as a continuum that exhibits certain plasticity characteristics.

3.1.1 Deviatoric (von Mises) and Hydrostatic Stresses and Strains

An important concept in stress analysis is that of separating a general stress state into the component that is tending to change the shape of the sample (i.e. the *deviatoric* component) and the one that is tending to change its volume (i.e. the *hydrostatic* component). As outlined in §2.3, this separation may be expressed as

$$\begin{bmatrix} \sigma_{11} & \sigma_{12} & \sigma_{13} \\ \sigma_{12} & \sigma_{22} & \sigma_{23} \\ \sigma_{13} & \sigma_{23} & \sigma_{33} \end{bmatrix} = \begin{bmatrix} \sigma_H & 0 & 0 \\ 0 & \sigma_H & 0 \\ 0 & 0 & \sigma_H \end{bmatrix} + \begin{bmatrix} \sigma_{11} - \sigma_H & \sigma_{12} & \sigma_{13} \\ \sigma_{12} & \sigma_{22} - \sigma_H & \sigma_{23} \\ \sigma_{13} & \sigma_{23} & \sigma_{33} - \sigma_H \end{bmatrix}$$

(3.1)

where the hydrostatic stress is a scalar given by

$$\sigma_H = \frac{1}{3}(\sigma_{11} + \sigma_{22} + \sigma_{33})$$

It may be noted that, since the sum of the diagonal terms is an invariant (Eqn. (2.14)), the hydrostatic stress is also the mean of the three principal stresses. Of course, Eqn. (3.1) also holds when expressed in terms of principal stresses, so that

$$\begin{bmatrix} \sigma_1 & 0 & 0 \\ 0 & \sigma_2 & 0 \\ 0 & 0 & \sigma_3 \end{bmatrix} = \begin{bmatrix} \sigma_H & 0 & 0 \\ 0 & \sigma_H & 0 \\ 0 & 0 & \sigma_H \end{bmatrix} + \begin{bmatrix} \sigma_1 - \sigma_H & 0 & 0 \\ 0 & \sigma_2 - \sigma_H & 0 \\ 0 & 0 & \sigma_3 - \sigma_H \end{bmatrix}$$

(3.2)

and the term on the right represents the deviatoric component of the stress state.

The *yield stress*, σ_Y, of a metal is conventionally taken to be the applied uniaxial stress at the onset of plasticity. It is commonly treated as a scalar. In fact, plastic deformation of metals is stimulated solely by the deviatoric component of the stress state and is unaffected by the hydrostatic component. This is consistent with plastic deformation (of metals) occurring with *no volume change*. The *von Mises stress* is given by

$$\sigma_{vM} = \sqrt{\frac{(\sigma_1 - \sigma_2)^2 + (\sigma_2 - \sigma_3)^2 + (\sigma_3 - \sigma_1)^2}{2}}$$

(3.3)

where σ_1, σ_2 and σ_3 are the *principal stresses* (see §2.2.2). It can be seen that the von Mises stress is also a scalar quantity, which can only be positive.

Under simple uniaxial tension or compression, the magnitude of the von Mises stress is equal to that of the applied stress, while the hydrostatic stress is equal to one third of it. The von Mises stress is always positive, while the hydrostatic stress can be positive or negative. It's not appropriate to think of the von Mises stress as being "tensile," as one would if it were a normal stress (with a positive sign). It's effectively a type of (volume-averaged) shear stress. Shear stresses do not really have a sign, but it's conventional to treat them as positive, as indeed is done for the von Mises stress.

It is common to regard the von Mises stress and the deviatoric stress as referring to the same quantity, and indeed the terms are sometimes used interchangeably. Strictly speaking, they cannot in fact be the same, since the von Mises stress is a scalar, while the deviatoric component of the stress state is a (second rank) tensor. However, the von Mises stress does represent the overall driving force for a shape change (strain) and it is often very convenient to be able to use this in place of a full tensorial treatment. The same is true of the resultant strain. Deviatoric and hydrostatic components of the (plastic) strain state can also be identified. The **von Mises strain,** often termed the "*equivalent plastic strain,*" is given by an analogous equation to Eqn. (3.3). Again, it always has a positive sign, but this does not mean that it is a "tensile" strain. It is a scalar representation of the deviatoric component of the strain state. As shown below in §3.2, the plastic deformation of a material can be characterized using constitutive laws that relate the von Mises stress to the von Mises strain. The hydrostatic plastic strain, on the other hand, always has a value of zero. This follows from the fact that plastic strain does not involve a change in volume. (This is not true of elastic strains, which do in general involve a volume change.)

3.1.2 Yield Envelopes in Stress Space – von Mises and Tresca Criteria

This concept needs to be taken a little further in order to understand how yielding is treated when a sample is subjected to a general stress state. In (principal) stress space, there is a line along which $\sigma_1 = \sigma_2 = \sigma_3$; this is termed the **hydrostatic line**. All stress states lying along this line have no deviatoric component and hence cannot cause any plasticity. The deviatoric component of a stress state is represented by the distance in stress space from this line. A logical expectation for a yield envelope in stress space is therefore that it should have the form of a cylinder having this line as axis – see Fig. 3.1.

The radius of this cylinder is readily obtained, in terms of the uniaxial yield stress, σ_Y, by considering that case in stress space, as shown in Fig. 3.1. It should first be noted that the angle ϕ (between the hydrostatic line and each of the principal stress directions) is given by

$$\phi = \cos^{-1}\left(\frac{1}{\sqrt{3}}\right) \tag{3.4}$$

which has a value of about 54.7°. Geometrical relationships such as this are readily demonstrated. For example, those familiar with the basics of crystallography will know that, in a cubic system, the angle between the normals to planes defined by **Miller indices** (see §4.1.3) of (h_1, k_1, l_1) and (h_2, k_2, l_2) is given by

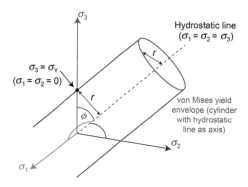

Fig. 3.1 The von Mises yield envelope in (principal) stress space, with a point marked that corresponds to yielding at an applied uniaxial stress of σ_Y.

$$\cos\phi = \frac{h_1 h_2 + k_1 k_2 + l_1 l_2}{\sqrt{h_1^2 + k_1^2 + l_1^2}\sqrt{h_2^2 + k_2^2 + l_2^2}} \quad (3.5)$$

Since the principal stress axes correspond to the normals to (100)-type planes and the hydrostatic line to the normal to the (111) plane, Eqn. (3.4) can be obtained from Eqn. (3.5).

It follows from this that, referring to the right-angled triangle in Fig. 3.1,

$$r = \sigma_Y \sin\phi = \sqrt{\frac{2}{3}}\sigma_Y \quad (3.6)$$

The envelope is thus defined in terms of the uniaxial yield stress (an experimentally obtained value). The equation for a cylinder with the hydrostatic line as axis may be written

$$(\sigma_1 - \sigma_2)^2 + (\sigma_2 - \sigma_3)^2 + (\sigma_3 - \sigma_1)^2 = \text{constant} \quad (3.7)$$

Substituting (for example) $\sigma_1 = \sigma_Y$ and $\sigma_2 = \sigma_3 = 0$ allows the constant to be evaluated as $2\sigma_Y^2$. The yield envelope is now defined by setting the von Mises stress (Eqn. (3.3)) equal to σ_Y. This allows assessment of whether any particular stress state (a set of σ_1, σ_2 and σ_3 values) will stimulate the onset of plasticity.

Two points should be noted at this stage. Firstly, while it certainly appears reasonable that the yield envelope should have this shape, the physical condition to which it corresponds is not immediately clear. In fact, the von Mises criterion is sometimes known as the ***maximum distortion energy criterion***, since it can be regarded as being related to the stored elastic strain energy due to deviatoric stresses at the point of yield. It is usually accepted that Richard Edler von Mises was the first to clearly formulate this criterion (in 1913), although others, including James Clerk Maxwell, did express similar ideas somewhat earlier. Mathematically, it corresponds to the second invariant of the stress state (Eqn. (2.14)) reaching a critical value.

3.1 The Onset of Plasticity and Yielding Criteria

There is, of course, quite extensive literature [1–6] that covers these issues in some detail, plus many more papers that relate to specific further conditions (such as material anisotropy, inhomogeneities, the presence of porosity etc.). However, the key message in terms of practical usage is simply that the von Mises stress is readily evaluated and that setting its value equal to the uniaxial yield stress is usually found to be a reliable way of predicting the onset of yielding. Furthermore, as outlined in the next section, the progression of work hardening after the onset of yielding can also be handled by simply monitoring the (rising) value of the von Mises stress.

The second point to note at this stage is that the von Mises criterion is not the only one for prediction of the onset of plasticity. In fact, several have been proposed, but the only other one that is really worthy of note (for metals) is that of **Tresca**. In terms of physical interpretation, the Tresca criterion is simpler than that of von Mises, since it states that yielding is expected when the peak shear stress in the system reaches a critical value. This has a logical basis, since dislocation glide (and also deformation twinning – see Chapter 4) are stimulated by shear stresses. The Tresca yield criterion is taken to be the work of Henri Tresca (1864), so it predates the von Mises criterion by about half a century. It does have a logical basis in terms of the micro-mechanisms of plastic deformation, although it takes no account of the normal stress on the plane of peak shear stress, which might be expected to have a (small) effect on the ease of dislocation glide.

Mohr's circle (§2.2.3) can be used to explore the Tresca yielding condition for a general stress state. This is illustrated in Fig. 3.2, where it can be seen that the magnitude of the peak shear stress is given by

$$\tau_{max} = \frac{\sigma_1 - \sigma_3}{2} \tag{3.8}$$

in which σ_1 and σ_3 are respectively the largest (most tensile) and smallest (most compressive) of the principal stresses. This shear stress acts in the 1–3 plane, at an angle of 45° (90° in Mohr space) to the 1 and 3 directions. Of course, with a

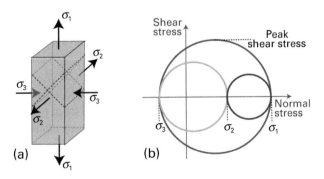

Fig. 3.2 Operation of a general set of principal stresses on a sample, showing (a) their orientation, with the planes marked on which the peak shear stresses operate, and (b) how the peak shear stress is obtained using the Mohr's circle construction.

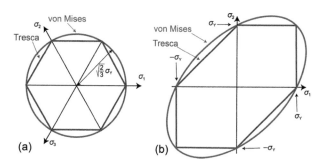

Fig. 3.3 Sections through (principal) stress space containing both Tresca and von Mises yield envelopes, showing (a) a projection along the hydrostatic line and (b) a plane stress section (for $\sigma_3 = 0$).

hydrostatic stress state ($\sigma_1 = \sigma_2 = \sigma_3$), all of the Mohr's circles are simply points, there are no shear stresses anywhere and thus no possibility of yielding.

This criterion is thus simple and intuitive, although its implementation does require inspection of the three principal stress values (to identify the largest and smallest), whereas this is not necessary with the von Mises criterion – in that case all three of the principal stresses are simply put into Eqn. (3.3). The Tresca criterion can also be plotted (as a yield envelope) in stress space. It turns out to be an hexagonal prism, with the hydrostatic line as axis, inscribed within the von Mises cylinder. This is shown in Fig. 3.3, plotted firstly in projection along the hydrostatic line (sometimes termed the π-plane) and secondly in section for a plane stress condition ($\sigma_3 = 0$).

The shapes of the yield envelopes (for plane stress) in Fig. 3.3(b) can readily be understood. In the positive and negative quadrant, the Tresca shape is clearly a square of side σ_Y (with σ_3 being the "other" stress in all cases). In the other two quadrants, it is σ_1 and σ_2 that provide both the "most tensile" and the "most compressive" stresses, giving the linear plots shown. The von Mises criterion (Eqn. (3.7)) reduces in the plane stress case to

$$\sigma_Y^2 = \sigma_1^2 + \sigma_2^2 - \sigma_1\sigma_2 \tag{3.9}$$

This is an equation for an ellipse, which passes through the points (σ_Y, 0), (σ_Y, σ_Y), (0, σ_Y), ($-\sigma_Y$, 0), ($-\sigma_Y$, $-\sigma_Y$) and (0, $-\sigma_Y$). The Tresca envelope also passes through these points. Furthermore, the ellipse also passes through the points ($-\sigma_Y/\sqrt{3}$, $\sigma_Y/\sqrt{3}$) and ($\sigma_Y/\sqrt{3}$, $-\sigma_Y/\sqrt{3}$). It follows that the ellipse has an aspect ratio of $\sqrt{3}$.

A quick summary regarding these two yield envelopes should include the following points. Firstly, the von Mises criterion is usually found to be more reliable. Secondly, it is in many ways easier to use, since it can be implemented by simply inputting the three principal stresses, without needing to check which pair gives the highest shear stress. Thirdly, it is more "conservative," since the Tresca envelope always lies inside it. Finally, use of the von Mises stress is well suited to treatment of the post-yielding (work hardening) behavior, as described in §3.3.4. Nevertheless, familiarity with the Tresca criterion is worthwhile, partly because it is closely related to the physical phenomena that take place during metal plasticity.

3.1.3 True and Nominal (Engineering) Stresses and Strains

Before moving on to treat the progressive plasticity that commonly follows the onset of yielding, it's important to have an appreciation of the nature of the actual (true) values of the stresses and strains once the (plastic) strain starts to become substantial (i.e. more than a few %). As was noted in §1.3, elastic strains rarely exceed about 1% (except with rubbers). At these levels, the difference between nominal and true stresses can usually be ignored. Plastic strains, however, often reach several tens of % and at these levels the differences become highly significant. Similar arguments apply to nominal and true strains.

For example, it is common practice during uniaxial (tensile or compressive) testing to equate the stress to the force divided by the original sectional area and the strain to the change in length (along the loading direction) divided by the original length. In fact, these are **"engineering"** or **"nominal"** values. The **true stress** acting on the material is the force divided by the current sectional area. After a finite (plastic) strain, under tensile loading, this area is less than the original area, as a result of the lateral contraction needed to conserve volume, so that the true stress is greater than the nominal stress. Conversely, under compressive loading, the true stress is less than the nominal stress.

Consider a sample of initial length L_0, with an initial sectional area A_0. For an applied force F and a current sectional area A, conserving volume, the true stress can be written

$$\sigma_T = \frac{F}{A} = \frac{FL}{A_0 L_0} = \frac{F}{A_0}(1 + \varepsilon_N) = \sigma_N(1 + \varepsilon_N) \tag{3.10}$$

where σ_N is the nominal stress and ε_N is the nominal strain. Similarly, the true strain can be written

$$\varepsilon_T = \int_{L_0}^{L} \frac{dL}{L} = \ln\left(\frac{L}{L_0}\right) = \ln(1 + \varepsilon_N) \tag{3.11}$$

The true strain is therefore less than the nominal strain under tensile loading, but has a larger magnitude in compression. While nominal stress and strain values are often plotted for uniaxial loading, it is essential to use true stress and true strain values throughout when treating more general and complex loading situations. These are usually von Mises stresses and strains (§3.3.2).

Figure 3.4 illustrates typical differences that can arise between true and nominal values. This is a stress–strain plot for a material that exhibits **linear work hardening**. (As will be seen in §3.3.4, this is not so common for metals, but it simplifies the treatment for current purposes). Such a stress–strain relationship may be written

$$\sigma = \sigma_Y + K\varepsilon \tag{3.12}$$

where σ_Y is the (true) yield stress, ε is the **true plastic strain** and K is the **work hardening coefficient**. The figure shows both true stress–true strain and nominal

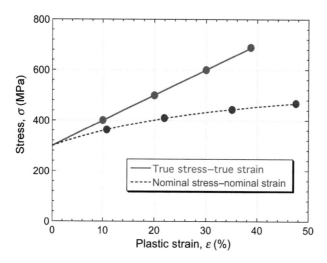

Fig. 3.4 Stress–strain plots, in true and nominal forms, for a metal with a yield stress of 300 MPa and a (linear) work hardening coefficient, K, of 1,000 MPa.

stress–nominal strain plots, for specific values of σ_Y and K. The true plot simply corresponds to Eqn. (3.12), while the nominal plot was obtained from it using Eqns. (3.10) and (3.11). It is clear that, for strains beyond a few % or so, the differences between the two plots become substantial, both for the values of stress and strain at any particular point and in terms of the overall shapes of the two plots. This is one of a number of important points relating to interpretation of a (nominal) stress–strain plot obtained under uniaxial loading. Further details are provided in Chapters 5 and 6.

3.2 Constitutive Laws for Progressive Plastic Deformation

3.2.1 Quasi-Static Plasticity – the Ludwik–Hollomon and Voce Laws

There is, of course, keen interest in how plastic deformation develops after its onset (yielding). It should first be noted that, from a mechanistic point of view, this is a very complex area. The mechanisms involved, and their dependence on microstructure, temperature, strain rate etc., are described in Chapter 4. All that needs to be noted at this point is that there is a marked tendency for the necessary (deviatoric) stress (sometimes termed the "*flow stress*") to rise as the plastic strain increases. However, the **work hardening rate** (gradient of the stress–strain curve after yielding) can vary from virtually zero (an "elastic–perfectly plastic" metal) to a value that is not much less than the gradient in the elastic portion – i.e. the Young's modulus. Also, the way that the work hardening rate changes (usually reduces) with increasing plastic strain can vary substantially.

Therefore, while the yield stress is usually taken to have a single value, work hardening needs more complex definition if it is to be represented analytically – i.e. to

3.2 Constitutive Laws for Plastic Deformation

be captured in a *constitutive law*. This should ideally be valid over an appreciable range of plastic strain – perhaps 50% or more in some cases. Even metals that are relatively hard (and brittle) are normally required to have *ductility* levels (plastic strains to failure) of at least several % if they are to be used for engineering purposes. It is important to note that constitutive laws are normally defined as a true stress–true strain relationship. It can be seen from Fig. 3.4 that the same relationship will tend to look very different when plotted as nominal stress against nominal strain.

Of course, there is no expectation that the work hardening curve will in fact conform to any particular functional form. However, in general, the work hardening rate (gradient of the true stress–true strain plot) tends to decrease progressively with increasing strain, perhaps eventually approaching a plateau. This is a consequence of competition between the creation of new dislocations, and inhibition of their mobility (by forming tangles etc.), and processes (such as climb and cross-slip) that will allow them to become more organized and to annihilate each other. (Further details are provided in Chapter 4.) Initially, the former group of processes tends to dominate, but a balance may eventually be reached, so that the "*flow stress*" ceases to rise – i.e. the work hardening rate becomes zero. (With metals, it is very rare, except with single crystals – see §4.2.1 and Fig. 4.11 – for the work hardening rate to rise with increasing strain, although this is quite common in certain types of polymer, as a consequence of molecular reorganization.)

Several analytical expressions have been proposed to characterize the work hardening of metals, but only two are in frequent use. The first is the **Ludwik–Hollomon** equation [7]

$$\sigma = \sigma_Y + K\varepsilon^n \quad (3.13)$$

where σ is the (von Mises) applied stress, σ_Y is its value at yield, ε is the plastic (von Mises) strain, K is the **work hardening coefficient** and n is the **work hardening exponent**. The second is the **Voce** equation [8]

$$\sigma = \sigma_s - (\sigma_s - \sigma_Y)e^{-\varepsilon/\varepsilon_0} \quad (3.14)$$

The stress σ_s is a **saturation level**, while ε_0 is a **characteristic strain** for the exponential approach of the stress towards this level. The range of shapes obtainable by varying the relevant parameters is illustrated by the plots in Fig. 3.5. It can be seen that a wide variety of curve shapes can be obtained.

This is the basis for a large global activity involving (FEM) simulation of the plastic deformation of metals, with the stresses and strains concerned being von Mises values. There is a strong incentive, for any particular metal, to identify the most appropriate constitutive law and the best-fit values of the parameters in it. A strong caveat should, however, be appended to such activities, since, even for a particular well-defined alloy, the plasticity characteristics can change substantially as a result of heat treatment, prior plastic deformation, exposure to radiation or chemically aggressive environments etc. The only way to be sure about how a particular material will behave when deformed plastically is to test it experimentally.

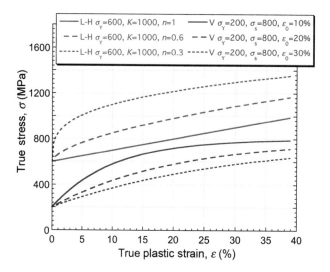

Fig. 3.5 True stress–strain curves in the form of Ludwik–Hollomon and Voce plots, obtained using Eqns. (3.13) and (3.14), with the parameter values shown in the legend.

It is naturally of considerable interest to explore how well actual experimental stress–strain data can be captured using these (or other) analytical expressions. In practice, this is not easy to illustrate in a compact way, since the number of such curves published over the past half-century or so is astronomical. Furthermore, there are a number of issues relating to the way that conventional (tensile) testing is carried out that can complicate such comparisons – see Chapter 5. In particular, there is something of a trend towards using relatively small test samples, which can create anomalies [9]. Nevertheless, most experimental curves can in practice be captured fairly well using one or both of the above two equations.

This is illustrated by Fig. 3.6, which compares experimental data [10] for three different types of stainless steel with best-fit plots obtained using Ludwik–Hollomon and Voce expressions. Several points may be noted here. Firstly, it can be seen that, despite the three experimental curves having very different shapes, all of them can be captured quite well using these simple analytical equations. Secondly, while this tends to be the case for the vast majority of tensile test curves, there is no expectation that the plots can be replicated exactly, particularly if they cover a large strain range. Finally, it should be emphasized that agreement can only be anticipated up to the point of *necking*, which is expected to start at the peak of the nominal stress–nominal strain curve – see §5.3.2. After that point, the distributions of strain and stress in the sample start to become highly inhomogeneous and their relationship with the plot becomes complex. It's no longer possible to convert between nominal and true stresses and strains using Eqns. (3.10) and (3.11). However, this complexity can in many cases be captured in an FEM model of the test, often by using a single true stress–true strain relationship and applying it so as to simulate neck formation and even the final fracture event (based on a critical strain criterion). This issue is fully covered in Chapter 5 (particularly §5.3.3).

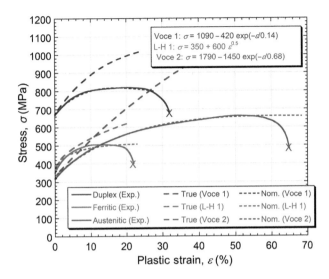

Fig. 3.6 Comparison between experimental tensile test data for three different types of stainless steel [10], reported as nominal stress–nominal strain plots, and best-fit curves obtained using Ludwik–Hollomon or Voce equations (for true stress–true strain relationships).

3.2.2 The Strain Rate Dependence of Plasticity and the Johnson–Cook Formulation

It is well known that metals exhibit strain rate-dependent plastic flow behavior. As a rule of thumb, the material response (stress–strain curve) starts to deviate significantly from the time-independent (*quasi-static*) version for strain rates above about 10^3 s^{-1}. The mechanisms responsible for this sensitivity (related to limits on dislocation mobility and alternative deformation modes) are known (and described in Chapter 4), but it cannot be predicted in any fundamental way and it is not easy to capture experimentally (for high strain rates). Nevertheless, simulation of this strain rate dependence is essential for FEM modeling [11] of many situations involving high strain rates (ballistics, explosions, crashes, machining, cutting etc.). Since such modeling is routinely carried out on a massive global scale, often with little scope for cross-checking of its reliability, validation of the procedures, and accurate quantification of the characteristics for a range of materials, are very important.

In order to implement such modeling, appropriate constitutive laws, and the values of parameters in them, are required. The most commonly used expression for high strain rate behavior, particularly for use in FEM models, is that of **Johnson and Cook** [12], which is often written in the form

$$\sigma = \left\{\sigma_Y + K\varepsilon_p^n\right\}\left[1 - \left(\frac{T - T_0}{T_m - T_0}\right)^m\right]\left(1 + C\ln\left\{\left(\frac{d\varepsilon_p}{dt}\right)\bigg/\left(\frac{d\varepsilon_p}{dt}\right)_0\right\}\right) \quad (3.15)$$

where σ_Y is the yield stress, K is the strain hardening coefficient, n is the strain hardening exponent, T_m is the melting temperature, T_0 is a reference (ambient) temperature, m is the temperature coefficient, $(d\varepsilon_p/dt)$ is the (plastic) strain rate, $(d\varepsilon_p/dt)_0$ is a reference (quasi-static) strain rate and C is the **strain rate sensitivity parameter**.

It can be seen that the first term is just the Ludwik–Hollomon expression (Eqn. (3.13)), which dictates the quasi-static yielding and work hardening behavior. (The Voce expression could alternatively be used.) The second term represents the temperature dependence and the third the strain rate sensitivity. This last term includes the logarithm of the strain rate, normalized by a reference value. Apart from this **normalizing strain rate**, usually taken to be a quasi-static rate, only the value of C is required in order to characterize the strain rate sensitivity. It should be noted, however, that the temperature dependence is often significant, since the imposition of rapid plastic straining is likely to cause local temperature rises, with most of the plastic work normally being released as heat – see §5.4.3.

It may, however, be noted that, for purposes of establishing the value of the strain rate sensitivity parameter, C, it's not essential to use an analytical expression for the first term in Eqn. (3.15). Since this $\sigma(\varepsilon)$ relationship is taken to be fixed, an experimentally obtained set of data pairs could be employed, instead of a functional relationship. On the other hand, for purposes of using the expression in FEM models, a fully analytical equation, such as Eqn. (3.15), is often more convenient.

Equation (3.15) is in practice considered to provide a fairly realistic representation of the behavior, at least in the regime below that in which shock waves [13] are likely to have a significant effect (i.e. it should be reliable for sub-sonic impact velocities). Several minor variations to the Johnson–Cook (J–C) formulation have been put forward [14–16], but the basic form is in general considered to be quite reliable. It should, however, be recognized that, as with all constitutive laws, the J–C formulation is essentially just an empirical expression and there have certainly been criticisms [17] of it. Nevertheless, it is in widespread use. Testing procedures aimed at experimental measurement of strain rate sensitivity parameters are covered in Chapter 10 (§10.4).

3.2.3 Constitutive Laws for Superelasticity

Superelasticity (SE) is the term commonly used to describe the generation of relatively large strains via a mechanism that involves a martensitic (diffusionless) phase transformation, which takes place via a shear mechanism. It is thus rather similar to plasticity induced by **deformation twinning**. (These mechanisms are described in more detail in Chapter 4.) However, under some circumstances, the phase transformation can subsequently be reversed, such that what appeared to be a plastic strain is recovered. (Such recovery is not normally possible with deformation twinning, or indeed with dislocation glide.) In fact, metals that can exhibit this behavior are sometimes termed **shape memory alloys** (SMAs), because they retain a "memory" of their shape before being mechanically deformed and can recover it by a suitable treatment (such as a temperature change). It should be noted, however, that under some conditions an alloy can exhibit SE, but not the shape memory effect (SME).

Some of the characteristics involved can be seen in Fig. 3.7. This shows experimental data [18] for a particular Ni-Ti alloy, in the form of (a) a DSC (*differential*

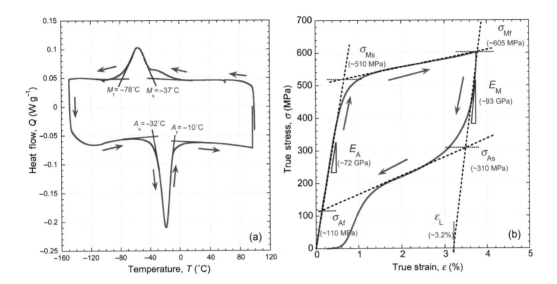

Fig. 3.7 Experimental data [18] for a Ni-Ti shape memory alloy, showing (a) a DSC plot giving the phase transformation temperatures during cooling and reheating and (b) a stress–strain curve, together with best-fit values of the parameters used to represent it.

scanning calorimetry) scan and (b) a stress–strain curve. The parent **austenitic** phase is stable at high temperature and the product **martensitic** phase is stable at lower temperature. However, it can be seen in the DSC plot that the system exhibits some **hysteresis**. During cooling, conversion to the martensitic phase is complete at the temperature M_f. On reheating, the reverse transformation back to the austenitic phase is complete at a (higher) temperature A_f. Above A_f, SE is possible, with large mechanically imposed strains (up to ~8%) being accommodated by transformation of the parent phase to metastable martensitic variants. These variants revert to the parent phase on removal of the applied load. Since A_f is about −10 °C, SE is exhibited at ambient temperature with this alloy.

The **shape memory effect** (SME) can also be observed in these alloys. Application of stress at a temperature below M_f can lead to the strain being accommodated by reorientation of martensitic variants. On heating above A_f, however, the martensite can transform to the parent phase in such a way that the original shape is recovered. Subsequent (unloaded) cooling below M_f can occur without further shape change. Repeated cycles of deformation, followed by heating to give shape recovery, are possible and other types of shape memory behavior can also be observed.

The stress–strain curve of Fig. 3.7(b) can thus be understood in terms of the mechanisms involved in the SE effect. The initial part of the loading curve is simply elastic deformation of the austenitic phase (with a well-defined Young's modulus value of about 72 GPa). When the stress reaches a certain value σ_{Ms} (~500 MPa in this case), what appears to be conventional plasticity is stimulated. However, this "yielding" and "plasticity" are actually taking place due to progressive austenite–martensite phase transformation. When this is complete, at a stress σ_{Mf} (~600 MPa), loading is reversed. (If the load were to be increased further, then conventional irreversible plasticity in the form of dislocation glide would take place.) Elastic unloading of the martensitic phase then occurs (with a different Young's modulus from that of the austenitic phase). This is followed, when the stress has fallen to σ_{As} (~300 MPa), by recovery of the "plastic" strain as the phase transformation is reversed. Eventually, as the sample becomes completely unloaded, virtually all of the strain is recovered (so that it can in this respect be regarded as having all been "elastic").

As can be seen in Fig. 3.7(b), the experimental curve is conventionally represented via the values of seven parameters, with linear plots assumed between the corresponding points in stress–strain space. The representation is thus a little more complex than that for conventional plasticity (requiring only three parameters for Ludik–Hollomon and Voce equations), although it should be noted that the Young's moduli are effectively pre-defined and also that the loading and unloading parts of the curve are more or less independent. On the other hand, it can be argued that even these seven parameters don't actually capture the behavior very well, particularly towards the end of the unloading. This issue is relevant to attempts to extract superelastic characteristics from indentation data, described in §8.4.5.

3.2.4 Progressive Creep Deformation and the Miller–Norton Law

Creep is a term commonly used to describe progressive (permanent) deformation of a material under load, usually over relatively long time scales. The mechanisms responsible can be relatively complex, but they usually involve some kind of thermally activated process, such as diffusion (or climb of dislocations). This commonly leads to a strong (***Arrhenius***) dependence of the strain rate on temperature. During loading under a constant stress, the strain tends to vary with time approximately as shown in Fig. 3.8, where the effect of changing the applied stress is also indicated. This is a plot of *creep strain*, not overall strain, against time, and hence the initial (elastic) strain at each stress has been omitted.

As shown in this figure, it is commonly observed that, after an initial transient (***"primary" creep***), the strain rate becomes stable for an extended period (***"secondary" creep***). Broadly, primary creep corresponds to the setting up of some kind of microstructural balance, which is then maintained during the quasi-steady state of secondary creep (before the onset of microstructural damage, such as void formation, when a final ***tertiary regime*** is entered). For example, this balance might be one in which dislocations are surmounting obstacles at a rate dictated by diffusional processes. More detail is provided in Chapter 4 (§4.5).

Interest commonly focuses on the strain rate during secondary creep, although properties such as the ***creep rupture strain*** may also be of interest. The steady state strain rate is usually expressed as

$$\frac{d\varepsilon_{cr}}{dt} = A\sigma^n e^{-Q/(RT)} \tag{3.16}$$

3.2 Constitutive Laws for Plastic Deformation

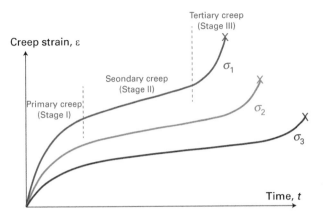

Fig. 3.8 Schematic creep strain curves, for three different stress levels ($\sigma_1 > \sigma_2 > \sigma_3$).

where A is a constant, σ is the applied stress, n is the stress exponent and Q is an activation energy. This is a ubiquitous formulation that is not normally associated with any particular name or names. It should be noted, however, that in practice primary creep sometimes extends over a substantial fraction of the period of interest. Also, the transition between primary and secondary regimes is often not very well defined and indeed it may be that a true steady state is never established. Formulations that cover both regimes can therefore be helpful. One example (often termed the **Miller–Norton law**) gives the creep strain as a function of time:

$$\varepsilon_{cr} = \frac{C\sigma^n t^{m+1}}{m+1} e^{-Q/(RT)} \qquad (3.17)$$

where C is a constant and m is a parameter controlling the way that the transition between the two regimes takes place. Plots showing the general shape of predicted curves, and the effects of changes in σ, n and m, are presented in Fig. 3.9. As expected, raising σ and, particularly, n, give sharp increases in strain rate. The sign of m is always negative: a lower magnitude delays the transition to the secondary (linear) regime, while a higher magnitude accelerates it (although, strictly, the strain rate never becomes exactly constant with this formulation).

Of course, there is also a high (exponential) sensitivity to temperature. As a very broad and general observation, creep tends to become significant only at temperatures above about $0.5\,T_m$ (i.e. 50% of the melting temperature in K) – i.e. at high **homologous temperatures**. For most metals, this translates into temperatures of at least a few hundred °C, although there are a few, such as Pb and Sn, for which ambient temperature is around $0.5\,T_m$. It should, however, again be appreciated that the equations used to describe creep are simply empirical relationships. While some sort of rationalization is sometimes possible – for example, the values of n and Q in Eqn. (3.17) may in some cases be correlated with identifiable atomic scale processes – these equations are not closely tied to microstructure or mechanisms. Of course, a similar statement can be made regarding the expressions used to describe plasticity

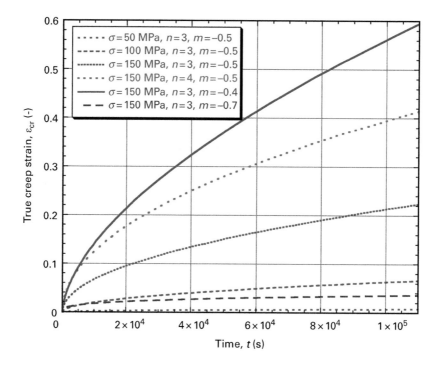

Fig. 3.9 Plots produced using the Miller–Norton law (Eqn. (3.17)), giving the creep strain as a function of time, for different values of the applied stress, σ, the stress exponent, n, and the parameter controlling the transition from primary to secondary regimes, m. The value of $C \exp(-Q/(RT))$ has been taken to be 10^{-10} MPa^{-n} s$^{-(m+1)}$.

(yielding and work hardening), although they can often be at least qualitatively interpreted in terms of microstructural effects such as dislocation interactions.

It may finally be noted that creep tests are commonly carried out with a constant applied load (rather than a constant true stress). Thus, for testing in tension, the tertiary regime may actually be a consequence of the fact that the true stress is rising throughout the test, with the rate of increase becoming greater towards the end of the test. At least in some cases, the tertiary regime may be associated with the true stress starting to approach or exceed the yield stress of the material. The situation can be further complicated by the possibility that significant microstructural changes (such as **recrystallization** – see §4.2.4), which could strongly affect the mechanical response, may occur during the test.

3.3 Residual Stresses

Metallic samples often incorporate residual stresses within them. Expressed as local forces, they must sum to zero when integrated over the whole sample, but locally they can reach quite high levels (at or above the yield stress of the material). They are thus inherently inhomogeneous – i.e. they must have different values (and signs) in different parts of the sample. They also commonly create **anisotropy** in the sample, so that it responds differently when loaded in different directions. They certainly have the potential to influence the outcome of mechanical testing procedures that are designed to investigate plasticity characteristics. They can also create **distortions** in the **shape** of the sample. Some fairly simple points are made here about how they can arise and about the types of effect that they can generate. More detailed treatment of specific cases is provided elsewhere, including some information about how residual stresses can be characterized via experimental measurements (§8.4.3).

3.3.1 Origins of Residual Stress

Many metallic components and samples contain residual stresses. For example, forged or rolled material has often undergone different degrees of plastic deformation in different parts. These parts are nevertheless constrained to be bonded together, which requires them to be internally stressed (and strained). This is an effect of very broad significance, closely related to the concept of a "*misfit strain*" – i.e. a difference between the (stress-free) dimensions of two constituents that are in fact constrained to be bonded together in some way. An example is provided by "*shot peening*," a process involving plastic deformation of a near-surface layer of a component, in such a way that it tends to become thinner and "wider." It remains, however, bonded to the underlying bulk. The final state is one in which a "*force balance*" has come into operation, such that the (plastically stretched) surface layer is forced back towards its original dimensions, creating a large (residual) compressive stress in all in-plane directions. In compensation, tensile stresses are created in the underlying bulk, although, since this is usually much thicker, those stresses tend to be relatively small. (A similar effect arises during cooling of a substrate/coating system, when the thermal expansion coefficient of the coating is less than that of the substrate.) In addition, a "*moment balance*" must also operate in such situations and, if the bulk underneath the shot-peened layer were relatively thin, then this would be noticeable in the form of the component becoming curved. Full details of how such force and moment balances operate in different situations are provided in Chapter 11 of [19].

The example of shot peening is a well-known one, with a simple geometry. The process is often employed mainly in order to induce these near-surface compressive stresses, which have a beneficial effect in inhibiting the formation of surface cracks, thus improving the fatigue resistance – see §10.3.3. There have been many studies [20–23] of the induced residual stresses in shot-peened components and their effects on damage development during service. Some FEM outcomes are shown in Fig. 3.10 from Hong et al. [21], referring to a material with a yield stress of 600 MPa and a linear work hardening rate of 800 MPa. These relate to outcomes of single impact events with rigid spheres, showing in-plane (radial) stresses immediately under the impact site, as a ratio to the yield stress. Of course, real situations involve multiple superimposed impacts at random locations, but outcomes of single, isolated impacts are qualitatively similar. It can be seen in Fig. 3.10(a) that these stresses are

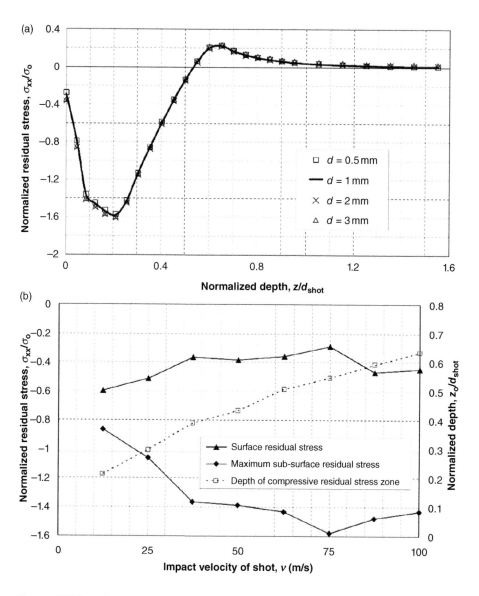

Fig. 3.10 FEM-predicted [21] outcomes of shot peening from single impact events, showing (a) the sub-surface distribution of the in-plane stress, for an impact velocity of 75 m s^{-1}, and (b) parameters concerning this distribution, as a function of the impact velocity.

compressive in a near-surface region, to a depth of approximately half of the diameter of the sphere. (Such stress fields are independent of the absolute scale, as is commonly the case.) At greater depths, this stress becomes tensile, although with smaller magnitudes. (A force balance does not appear to be satisfied in these profiles, but it would be seen to do so if the full 3-D stress field were to be examined.)

Of course, the depth of the compressive zone does depend on how deeply the sphere has penetrated, which in turn depends on the mass and impact velocity of the

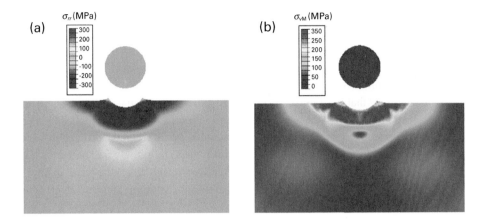

Fig. 3.11 FEM-predicted fields of: (a) the in-plane (radial) stress and (b) the von Mises stress, created by impact of a hard sphere with a velocity of 120 m s^{-1}, into a sample of as-received copper ($\sigma_Y \sim 350$ MPa, with little work hardening). Details of the boundary conditions used in the model are provided elsewhere [24]. The Johnson–Cook strain rate sensitivity parameter, C, was given a best-fit value of 0.016.

sphere and the properties of the material. The effect of shot velocity can be seen in Fig. 3.10(b), which shows that the magnitude of the peak (sub-surface) compressive stress goes up at higher impact speeds, while the value right at the surface drops slightly. As expected, the depth of the compressive region increases as the impact velocity is raised. A more complete picture of a typical (in-plane) stress field (arising from a single impact event) is provided by Fig. 3.11, which is from a model in which the effects of strain rate and (plasticity-induced) rises in temperature are taken into account [24], although these effects don't in general dramatically alter the nature of the stress and strain fields. The lower yield stress (and lower work hardening rate) of the (copper) material in Fig. 3.11, compared with that of Fig. 3.10, leads to lower stress magnitudes in the compressive zone, although it has a similar depth. The exact surface profile of the indent is quite sensitive to the work hardening characteristics – for example, see Fig. 8.3 in §8.2.2, but this is of little significance in terms of the (compressive) stress field created by shot peening. Overall, the main messages to note are simply that these near-surface compressive stresses are relatively high (of the order of the yield stress) and that they extend to a depth of the order of the shot radius – typically a mm or two.

Similar types of effect – i.e. differential plastic straining of different regions of a component during processing – arise in many forming procedures, such as *forging*, *deep drawing*, *rolling* etc. In fact, residual stresses arising from this source tend to be present in most metallic components, although their nature and significance varies over a wide range. Most such processes are more complex than shot peening, sometimes with local variations in temperature and/or strain rate during the operation being relevant to the outcome. Microstructural effects, such as recrystallization (§4.2.4) being induced during the process, can also be important.

It should also be recognized that residual stresses can arise from sources other than differential plastic straining – for example from temperature changes or phase changes (which have associated volume changes). **Welding** is a good example of a process in which there are substantial local changes in temperature and also phase changes (solidification). These also create **misfit strains** (between the **stress-free shapes** of adjoining regions that must in fact be bonded together), in a similar way to the creation of a misfit strain between near-surface and interior regions of a shot-peened component. **Castings** also tend to contain residual stresses and in fact it could be said that it is actually rather unusual for a metallic component or sample to be entirely free of them. Of course, their magnitude can vary substantially and in many cases they can safely be ignored. Nevertheless, the possible presence of residual stresses should always be borne in mind.

3.3.2 Effects of Residual Stress on Plasticity

The effect of a (well-characterized) residual stress field in a sample on its plasticity response during testing is actually quite easy to predict. The stress field from the applied load can simply be superimposed on the residual stress field, so that the net stress field at any point can be predicted, as can the way that plasticity is expected to proceed. In practice, even if the residual stress field is known, it does introduce complications, since it is likely to be inhomogeneous. Plasticity is thus expected to start in some locations earlier than in others, although of course, if the sample is to retain its coherence, this will often lead to a rather complex development of the stress and strain fields (and the dimensions of the sample).

This kind of case often can be analyzed, although it is likely to require numerical (FEM) modeling. Also, even if this tool is available, inferring what is actually happening inside a sample from measurable outcomes (such as surface strain data or changes in sample shape) might require iterative modeling (with trial and then improved values for plasticity parameters, or for an inferred residual stress field). These are often termed *inverse* FEM problems and they can be challenging. They figure quite strongly in later parts of this book and examples of how residual stress levels can be inferred in this way, from indentation data, are presented in §8.4.3.

References

1. Hosford, WF, Generalized isotropic yield criterion. *Journal of Applied Mechanics*, 1972. **39**(2): 607–609.
2. Yang, WH, A generalized Von Mises criterion for yield and fracture. *Journal of Applied Mechanics: Transactions of the ASME*, 1980. **47**(2): 297–300.
3. Goo, E and KT Park, Application of the Von Mises criterion to deformation twinning. *Scripta Metallurgica*, 1989. **23**(7): 1053–1056.
4. Capsoni, A and L Corradi, Variational formulations for the plane strain elastic–plastic problem for materials governed by the Von Mises criterion. *International Journal of Plasticity*, 1996. **12**(4): 547–560.

5. Ponter, ARS and M Engelhardt, Shakedown limits for a general yield condition: implementation and application for a Von Mises yield condition. *European Journal of Mechanics A: Solids*, 2000. **19**(3): 423–445.
6. Lagioia, R and A Panteghini, On the existence of a unique class of yield and failure criteria comprising Tresca, Von Mises, Drucker–Prager, Mohr–Coulomb, Galileo–Rankine, Matsuoka–Nakai and Lade–Duncan. *Proceedings of the Royal Society A: Mathematical Physical and Engineering Sciences*, 2016. 472(2185).
7. Hollomon, JH, Tensile deformation. *Transactions of the American Institute of Mining and Metallurgical Engineers*, 1945. **162**: 268–290.
8. Voce, E, The relationship between stress and strain for homogeneous deformation. *Journal of the Institute of Metals*, 1948. **74**(11): 537–562.
9. Zhao, YH, YZ Guo, Q Wei, TD Topping, AM Dangelewicz, YT Zhu, TG Langdon and EJ Lavernia, Influence of specimen dimensions and strain measurement methods on tensile stress-strain curves. *Materials Science and Engineering A: Structural Materials Properties Microstructure and Processing*, 2009. **525**(1–2): 68–77.
10. Arrayago, I, E Real and L Gardner, Description of stress–strain curves for stainless steel alloys. *Materials & Design*, 2015. **87**: 540–552.
11. Umbrello, D, R M'Saoubi and JC Outeiro, The influence of Johnson–Cook material constants on finite element simulation of machining of AISI 316L steel. *International Journal of Machine Tools & Manufacture*, 2007. **47**(3–4): 462–470.
12. Johnson, GR and WH Cook, A constitutive model and data for metals subjected to large strains, high strain rates and high temperatures, in *Proceedings of the 7th International Symposium on Ballistics*, 1983. 21: 541–547.
13. Molinari, A and G Ravichandran, Fundamental structure of steady plastic shock waves in metals. *Journal of Applied Physics*, 2004. **95**(4): 1718–1732.
14. Rule, WK and SE Jones, A revised form for the Johnson–Cook strength model. *International Journal of Impact Engineering*, 1998. **21**(8): 609–624.
15. Lin, YC, XM Chen and G Liu, A modified Johnson–Cook model for tensile behaviors of typical high-strength alloy steel. *Materials Science and Engineering A: Structural Materials Properties Microstructure and Processing*, 2010. **527**(26): 6980–6986.
16. He, A, GL Xie, HL Zhang and XT Wang, A comparative study on Johnson–Cook, modified Johnson–Cook and Arrhenius-type constitutive models to predict the high temperature flow stress in 20CrMo alloy steel. *Materials & Design*, 2013. **52**: 677–685.
17. Manes, A, L Peroni, M Scapin and M Giglio, Analysis of strain rate behavior of an Al 6061 T6 alloy, in *11th International Conference on the Mechanical Behavior of Materials*, Guagliano, M and L Vergani, eds. Amsterdam: Elsevier, 2011, pp. 3477–3482.
18. Roberto-Pereira, FF, JE Campbell, J Dean and TW Clyne, Extraction of superelasticity parameter values from instrumented indentation via iterative FEM modelling. *Mechanics of Materials*, 2019. **134**: 143–152.
19. Clyne, TW and D Hull, *An Introduction to Composite Materials*. 3rd ed. Cambridge: Cambridge University Press, 2019.
20. Meguid, SA, G Shagal, JC Stranart and J Daly, Three-dimensional dynamic finite element analysis of shot-peening induced residual stresses. *Finite Elements in Analysis and Design*, 1999. **31**(3): 179–191.
21. Hong, T, JY Ooi and B Shaw, A numerical simulation to relate the shot peening parameters to the induced residual stresses. *Engineering Failure Analysis*, 2008. **15**(8): 1097–1110.

22. You, C, M Achintha, KA Soady, N Smyth, ME Fitzpatrick and PAS Reed, Low cycle fatigue life prediction in shot-peened components of different geometries, part I: residual stress relaxation. *Fatigue & Fracture of Engineering Materials & Structures*, 2017. **40**(5): 761–775.
23. Guan, J, LQ Wang, YZ Mao, XJ Shi, XX Ma and B Hu, A continuum damage mechanics based approach to damage evolution of M50 bearing steel considering residual stress induced by shot peening. *Tribology International*, 2018. **126**: 218–228.
24. Burley, M, JE Campbell, J Dean and TW Clyne, Johnson–Cook parameter evaluation from ballistic impact data via iterative FEM modelling. *International Journal of Impact Engineering*, 2018. **112**: 180–192.

4 Mechanisms of Plastic Deformation in Metals

The capacity of metals to undergo large plastic strains (without fracturing) is one of their most important characteristics. It allows them to be formed into complex shapes. It also means that a component under mechanical load is likely to experience some (local) plasticity, rather than starting to crack or exhibit other kinds of damage that could impair its function. Metals are in general superior to other types of material in this respect. This has been known for millennia, but the reasons behind it, and the mechanisms involved in metal plasticity, only started to become clear less than a century ago and have been understood in real depth for just a few decades. Central to this understanding is the atomic scale structure of dislocations, and the ways in which they can move so as to cause plastic deformation, although there are also several other plasticity mechanisms that can be activated under certain circumstances. These are described in this chapter, together with information about how they tend to be affected by the metal microstructure. This term encompasses a complex range of features, including crystal structure, grain size, texture, alloying additions, impurities, phase constitution etc.

4.1 Basic Dislocation Structures and Motions

4.1.1 The Atomic Scale Structure of Metals

Dislocations can exist in various types of (crystalline) material, but they tend to be both more numerous and more mobile in metals, compared with, say, ceramics, intermetallics or organic materials. The underlying reason for this is related to the types of interatomic bonding in different materials. In non-metallic materials such as ceramics, there are strong directional bonds between neighboring atoms, usually either **ionic** (adjacent atoms carrying charge of opposite sign) or **covalent** (electrons localized in orbitals shared between neighboring atoms), or possibly there is an element of both types. This militates against the formation of dislocations (which are local disruptions to the atomic scale structure of the crystal), and also inhibits their motion (which requires the breaking and reforming of interatomic bonds).

Metals, on the other hand, have a different type of bonding (usually termed **metallic**!). They contain **delocalized electrons**, which are free to move through the assembly of atoms (creating both electrical and thermal conductivity). The individual atoms therefore carry a positive charge (and so can be regarded as ions), although of course they are surrounded by a "sea" of electrons and the material as a whole is electrically neutral. The interatomic forces therefore have the nature of an overall "cohesion," but without any strongly directional forces between adjacent atoms (ions). Under these circumstances, the energy barrier to formation of a dislocation, and also to its motion, is considerably less than in a material with directional interatomic forces. Also, the atomic scale disruption associated with a dislocation tends to extend over a considerably greater distance (perhaps 10 or 20 atomic diameters, compared with just a few diameters in other materials). This also facilitates their motion.

4.1.2 Concept of a Dislocation

While elastic deformation (of metals) simply involves small perturbations of the interatomic distances, it is clear that plastic deformation must involve some kind of permanent changes to the locations of atoms, with respect to others. It's also clear that this must arise from the deviatoric (shape-changing) component of the stress state – i.e. from shear stresses (see §3.1.1). It was originally thought that this must take place by sliding of atomic planes past each other, presumably oriented such that it occurred on planes experiencing peak shear stresses (e.g. at 45° to an applied uniaxial stress – see §3.1.2). However, estimates of the shear stress needed to do this, which can be made from the value of the shear modulus, G, gave values that were much larger than those observed experimentally – i.e. plastic deformation by sliding of atomic planes under shear takes place much more easily than expected, at least for relatively pure single crystals. The value expected for sliding of a complete plane is about $0.1\ G$, whereas observed values were below this by up to two orders of magnitude.

The explanation for this is that, when two atomic planes in a metal slide over each other under the action of a shear stress, they do not slip as if they were rigid sheets. In fact, slip occurs more easily if, instead of a complete plane slipping over another in a single operation, the slippage is *localized* and this location moves progressively across the plane. These localized regions are ***dislocations***. A simple representation of one is shown in Fig. 4.1 (although it should be noted that, as mentioned in §4.1.1, in practice the structural disruption often extends over rather more than the four or five atomic diameters shown in this figure).

Historical development of improved understanding of the mechanical behavior of metals is closely linked to dislocations. Landmark advances include the theoretical shear strength calculations of Jacov Frenkel (1926), the postulated structure of (edge) dislocations, due to Egon Orowan, Michael Polanyi and Geoffrey Taylor (all 1934), the structure of the screw dislocation, suggested by Johannes Burgers in 1939, and proposals regarding the formation and multiplication of dislocations by Charles Frank and Thornton Read in 1950. They were first observed directly (in the transmission electron microscope, or TEM) by Peter Hirsch in 1956. Historical reviews are available [1, 2] that summarize these and other developments.

Dislocations are ***line defects***. The one shown in Fig. 4.1 is an ***edge dislocation***, which can be thought of as an ***extra half-plane*** of atoms. ***Slip*** of a dislocation ("***glide***") is illustrated in Fig. 4.2, together with the analogy of the motion of a

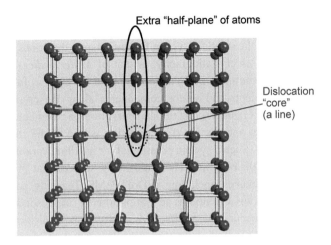

Fig. 4.1 Schematic representation of an edge dislocation (in a simple cubic structure).

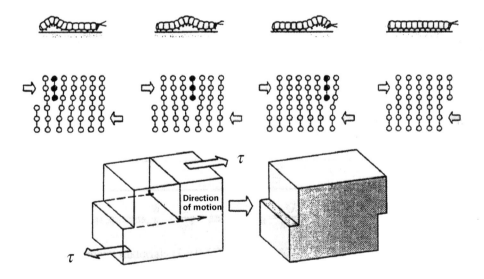

Fig. 4.2 Propagation of an edge dislocation across a plane, under the action of a shear stress.

caterpillar. (A better analogy is moving a carpet over a floor by propagating a ruck across it, rather than dragging the whole carpet, since the ruck is a "line defect" moving across a plane, like a dislocation, whereas the caterpillar propagates a "point defect" along its length.) The shear stress needed to cause a dislocation to move in this way is often called the **Peierls–Nabarro stress** (or just the **Peierls stress**), and it is sometimes referred to as the *lattice friction stress*. Detailed analysis of the interatomic forces involved [3] has confirmed that it is typically predicted to be only a very small fraction of the shear modulus (as observed experimentally).

4.1.3 The Dislocation Line Vector and the Burgers Vector

It should first be noted that the standard system for designating directions and planes, particularly in the context of crystallography, is that of **Miller indices**. These are numerical values used in denoting a plane as $(h\ k\ l)$ and a direction as $[u\ v\ w]$. They relate respectively to intercepts and vector components along the crystallographic axes. For example, the (1 1 1) plane is one of the (four) close-packed planes in a *face-centered cubic* (fcc) system and the [1 –1 0] direction is one of the (three) close-packed directions lying in it. Such a system is essential for handling various aspects of crystallography, including dislocations. There are many sources that provide basic information about Miller indices, including a DoITPoMS Teaching and Learning Package (TLP) available at www.doitpoms.ac.uk/tlplib/miller_indices/index.php.

Distortion of the lattice occurs along a line, which, for an edge dislocation, is the bottom row of the extra half plane of atoms (Fig. 4.1). This direction, characterized by a unit vector (the **dislocation line vector**), is an important characteristic of the dislocation, although it can change as the dislocation moves (see below). The other important characteristic is the **displacement** generated by its passage, which has both magnitude and direction. This displacement is the **Burgers vector**, **b**. It can be obtained by tracing a path (the **Burgers circuit** – see Fig. 4.3) around the dislocation. In a perfect crystal, it returns to the starting point. The dislocation causes a *closure failure*, the magnitude and direction of which define **b**.

For an edge dislocation, **b** is normal to the dislocation line vector. These two vectors define the **slip plane**. Slip occurs by the dislocation line moving (gliding) across the slip plane, generating a displacement **b** as it passes. The combination of the slip plane and the slip direction (direction of the Burgers vector) is often called the **slip system**. Certain types of plane and direction (commonly the most closely packed planes and directions) usually constitute the normal slip system in a given crystal structure. A possible combination of a slip direction and a slip plane in an fcc (face-centered cubic) structure are shown in Fig. 4.4, superimposed on the unit cell (with a corner atom removed).

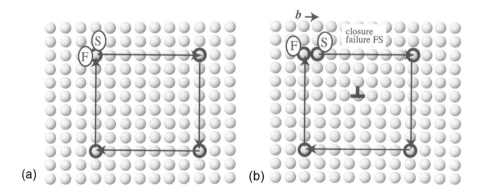

Fig. 4.3 A Burgers circuit (a) in a perfect crystal and (b) around an edge dislocation, with the core marked by the dislocation symbol, showing how the Burgers vector is obtained.

4.1.4 Screw Dislocations

Not all dislocations have *b* normal to their line vector. The complementary limiting case to an edge dislocation is one in which *b* is **parallel** to the line vector. This is termed a *screw dislocation*, the terminology arising from the helical displacement of the atoms around the core – see Fig. 4.5. The arrow indicates the direction of both the dislocation line and the Burgers vector. Since *b* and the line vector are parallel for a screw, there is a *family* of possible slip planes.

During slip (Fig. 4.6), the dislocation line moves normal to the applied shear stress (generating slip in the direction of the stress). However, dislocations cannot be immutably labelled "edge" or "screw." Most dislocations are "*mixed*" (line vector, l, at an angle between 0° and 90° to *b*). Moreover, since dislocation lines are rarely straight, their (edge/screw) nature changes along their length (and perhaps also during motion, when the shape may change). However, the Burgers vector, *b*, of a dislocation *is* immutable. When a dislocation moves, it always generates the same displacement of

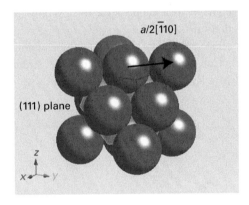

Fig. 4.4 Slip planes and directions in an fcc crystal.

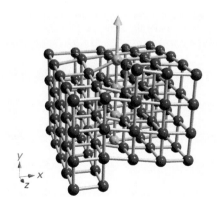

Fig. 4.5 Schematic representation of a screw dislocation (in a simple cubic structure).

the slipped part, relative to the unslipped region. In order for the crystal structure to be left unaffected by the passage of a dislocation, **b** must be a *lattice vector*; such dislocations are termed "*perfect*." This does not, however, apply to all dislocations. For example, "*partial*" dislocations, with Burgers vectors that are not lattice vectors, are described in §4.2.3.

4.1.5 Force on a Dislocation and Energies Involved

Consider the work done when a dislocation glides across a slip plane – see Fig. 4.7. This can be written both in terms of the total force on the dislocation line ($= FL$), acting over a distance d, and also in terms of the externally applied load ($= \tau A$, where A is the area of the slip plane), which generates a displacement b. It follows that

$$\tau A b = FLd = FA$$
$$\therefore F = \tau b \tag{4.1}$$

This applies to all types of dislocation. It should be noted that the expression for the force is actually a *dot product* (so that the magnitude of the force is only τb if the shear stress τ acts parallel to the direction of b).

The energy associated with a dislocation can be estimated by examining how the surrounding lattice is distorted. Consider an annulus of material around a screw dislocation – see Fig. 4.8. The shear strain in the annulus is given by

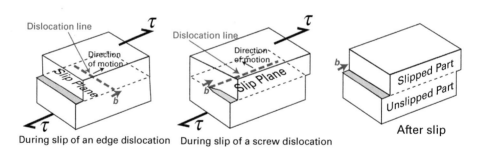

Fig. 4.6 Glide of edge and screw dislocations under the influence of a shear stress.

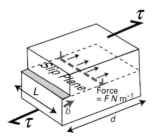

Fig. 4.7 Gliding of an edge dislocation.

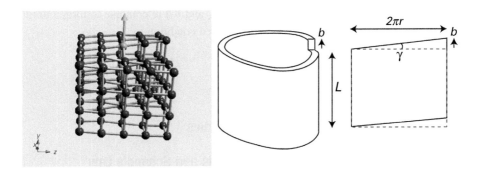

Fig. 4.8 Distortion of the lattice around a screw dislocation.

$$\gamma = \frac{b}{2\pi r} \quad (4.2)$$

The elastic strain energy per unit volume is given by the area under the stress–strain plot – i.e. ½(stress × strain). Therefore, in this case, the energy per unit volume is given by

$$U = \frac{1}{2}(G\gamma)\gamma \quad (4.3)$$

where G is the shear modulus. Since the volume of the annulus is $(2\pi rL\, dr)$, the energy stored in it is given by

$$dW = \frac{2\pi r\, dr\, LG\gamma^2}{2} = G\gamma^2\, \pi rL\, dr \quad (4.4)$$

The total energy is found by integrating between the core radius, r_0 and some outer limit, r_∞:

$$W = \int_{r_0}^{r_\infty} G\left(\frac{b}{2\pi r}\right)^2 \pi rL\, dr + \text{core energy}$$

$$\therefore W = \frac{Gb^2 L}{4\pi} \ln\left(\frac{r_\infty}{r_0}\right) + \text{core energy} \quad (4.5)$$

The integration limits are somewhat arbitrary, and calculation of the core energy is a little complex, but the result is not very sensitive to the exact assumptions and the outcome is that

$$W \sim \frac{Gb^2 L}{2} \quad (4.6)$$

This energy is usually expressed per unit length of dislocation line (i.e. a **line tension**):

$$\Lambda \sim \frac{Gb^2}{2} \quad (4.7)$$

The fact that dislocation energy is proportional to b^2 leads to a simple method, often called **Frank's rule**, for determining whether it is energetically favorable for two dislocations (with Burgers vectors b_1 and b_2) to combine to form a third dislocation, with a Burgers vector b_3. The required condition is

$$\left(b_1^2 + b_2^2\right) > b_3^2 \tag{4.8}$$

4.2 Further Dislocation Characteristics

4.2.1 Tensile Testing of Single Crystals and Schmid's Law

During tensile loading of a single crystal (Fig. 4.9), an applied stress will be reached when the shear stress, resolved onto a slip plane, in a slip direction, attains the critical level to cause dislocations in that slip plane to glide. It is called the *"primary"* slip plane, and the system that first comes into operation is the **primary slip system**.

If the normal *n* of this slip plane lies at an angle ϕ to the tensile axis, its area will be $A/\cos\phi$. Similarly, if the slip direction lies at an angle λ to the tensile axis, the component of the axial force F acting in the slip direction will be $F \cos \lambda$. The *resolved shear stress* is therefore given by

$$\tau_R = \frac{F \cos \lambda}{(A/\cos \phi)} = \sigma \cos \phi \cos \lambda \tag{4.9}$$

(Such resolving, i.e. transforming, of a stress, involving two cosine terms, is treated fully in §2.2.2.) For a given material, the value of τ_R at which slip occurs is usually found to be constant, often called the **critical resolved shear stress** (τ_{crit}). This is **Schmid's Law**, and $\cos\phi \cos\lambda$ is called the **Schmid factor** [4]. The applied stress at which slip starts is the yield stress, σ_Y. It follows that

$$\tau_{crit} = \sigma_Y \cos \phi \cos \lambda \tag{4.10}$$

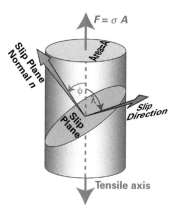

Fig. 4.9 Geometry of slip during tensile testing of a single crystal.

4.2 Further Dislocation Characteristics

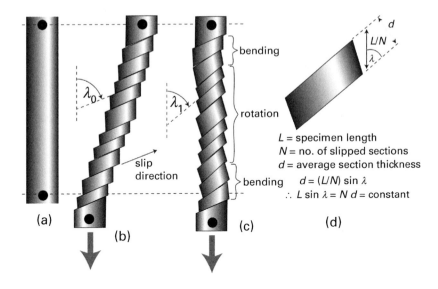

Fig. 4.10 Specimen reorientation during glide with a single slip system, (a) before loading, (b) with no lateral constraint, (c) with aligned grips and (d) section dimensions.

The Schmid factor is readily evaluated, for a given slip system (expressed using **Miller indices**) and orientation of the tensile axis. For example, in a cubic system, the expression presented as Eqn. (3.5) in §3.1.2 can be used. The maximum value of a Schmid factor is 0.5 (with both slip direction and slip plane normal inclined at 45° to the tensile axis. This is the case depicted in Fig. 3.2(a) in §3.1.2 (where the material is treated as an isotropic continuum, so that there are no constraints on the orientation of slip planes and slip directions). An example calculation of a Schmid factor is given in §9.5.3.

Slip on a single system tends to cause lateral displacement (as well as axial extension): if the grips are aligned so as to inhibit this, then the tensile axis rotates towards the slip direction (Fig. 4.10). Ignoring the short regions in which bending occurs, the specimen length, L, and inclination of the slip direction to the tensile axis, λ, are related (Fig. 4.10(d)) by

$$L_0 \sin \lambda_0 = L_1 \sin \lambda_1 \tag{4.11}$$

This reorientation will eventually cause other slip systems to operate – i.e. their Schmid factors to reach the same value as that of the primary system. This reorientation of the tensile axis (relative to the crystallographic axes), and the consequent activation of a second slip system at some stage, has a strong effect on the (single crystal) stress–strain curve. This is illustrated in Fig. 4.11.

4.2.2 Dislocation Interactions and Work Hardening Effects

The dislocation density, ρ, is a measure of the total dislocation line length per unit volume or, equivalently, the number of dislocations intersecting unit area. It has units of m^{-2}. The average spacing between dislocations, L, is related to the dislocation density:

$$L = \frac{1}{\sqrt{\rho}} \qquad (4.12)$$

Dislocation densities in metals range from ~10^{10} m^{-2} (annealed) to ~10^{16} m^{-2} (cold worked). Corresponding spacings are 10 μm to 10 nm. (This could lead to very small tested regions being free of dislocations, as described in §9.4.2.) It might be imagined that having more dislocations would facilitate plastic deformation, but in fact the opposite tends to occur. Plastic deformation becomes harder as ρ rises and gliding becomes inhibited by interactions, as described below.

There is an *elastic strain field* around a dislocation, where the lattice is distorted. This has stored energy (Eqn. (4.7)) and can create forces between dislocations, which tend to move so as to reduce these strains. Consider the strain fields of a pair of parallel edge dislocations, as shown in Fig. 4.12. If the dislocations are at a similar height

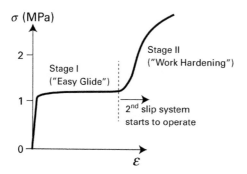

Fig. 4.11 Schematic stress–strain curve for a (pure, metallic) single crystal, showing how the initial easy glide (single slip system) regime gives way to multiple slip at higher strains, with a consequent increase in the stress level necessary to cause continued straining.

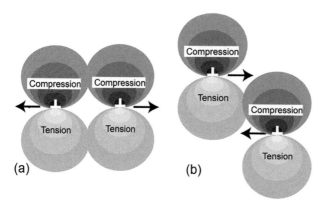

Fig. 4.12 Strain fields around pairs of parallel edge dislocations, and resulting glide forces acting on them, when they are (a) side-by-side and (b) at different heights.

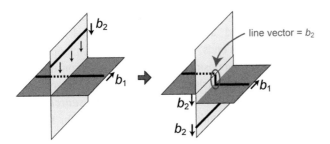

Fig. 4.13 Formation of jogs by intersection of two edge dislocations. The jog in b_2 is undetectable, whereas that in b_1 is sessile (unable to glide).

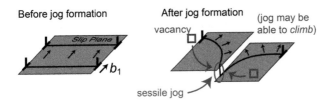

Fig. 4.14 Schematic depiction of how a sessile jog can reduce the mobility of a dislocation.

(Fig. 4.12(a)), then regions above the dislocation lines (under compression) will overlap, increasing the distortion. A similar effect occurs in regions of tension below the dislocation lines. The dislocations thus tend to glide away in opposite directions. If, on the other hand, the dislocations are at different heights, then they will tend to glide towards each other, so the compression region above one dislocation overlaps with the tension region below the other, thus reducing the strain energy.

Dislocations often meet other dislocations (of different types, on different slip planes), in various configurations. They may amalgamate to form new dislocations, annihilate each other, be repelled or cut through each other. When two dislocations intersect, *jogs* form in both. These are short sections with length and direction equal to **b** of the other dislocation. If a jog lies in a slip plane (when it is sometimes called a *kink*), it can glide with the rest of the dislocation, but if it lies out of the slip plane, it may be *sessile* (unable to glide, i.e. not *glissile*) – see Fig. 4.13. Such defects reduce dislocation mobility after straining, as shown in Fig. 4.14 (where the possibility of *climb* is noted – see below). Such defects and entanglements create work hardening – i.e. a requirement for higher applied stresses in order to allow straining to continue. The term "*forest hardening*" is also sometimes used. There is extensive literature [5–8] concerning the details of these interactions.

4.2.3 Creation of Dislocations and Types of Dislocation Mobility

Various effects can raise the dislocation density, ρ. Dislocations can be created at free surfaces, grain boundaries and within grains. As an example of the latter, consider a dislocation line fixed (pinned, e.g. by sessile jogs) at both ends, as shown in Fig. 4.15. When a shear stress τ is applied, a force $F = \tau b$ acts normal to the line. Since the dislocation line is pinned at both ends, it bows outwards into a loop, which eventually surrounds the pinning points. Opposing segments annihilate on meeting, forming a loop and recreating a short, pinned dislocation. This can repeat many times, generating a new loop each time. This is known as a **Frank–Read source**. There has been extensive work [9, 10] on the details of such mechanisms. In fact, as mentioned above, plastic straining can raise the dislocation density by several orders of magnitude, making individual dislocations much less mobile and raising the hardness considerably.

There is, however, scope for dislocations to move so as to reduce the incidence of such entanglements and sessile features. The ease with which this can occur has an effect on the work hardening rate. One of the mechanisms involved is **cross-slip**. If dislocations could only move by gliding on a single slip plane, their motion would be constrained and they would soon become impeded by obstacles, such as other dislocations or fine precipitates. However, screw dislocations can by-pass obstacles on the slip plane by **cross-slip** onto an alternative plane, as shown in Fig. 4.16. Figure 4.16(a) depicts a "family" of possible planes on which a pure screw dislocation (**b** // to line vector) can slip. By switching to an alternative one, the freedom of movement available to it is increased. Only a pure screw dislocation can do this, since, for other types of dislocation, the two vectors define a unique slip plane. However, a dislocation can often become pure screw, at least temporarily while it cross-slips, by locally aligning its line vector // to **b**.

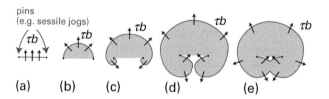

Fig. 4.15 The Frank–Read source.

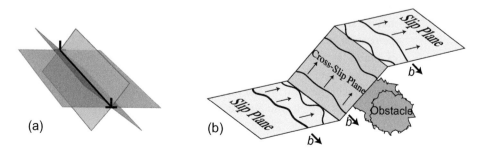

Fig. 4.16 (a) A family of possible slip planes available to a pure screw dislocation and (b) a dislocation undergoing cross-slip in order to avoid an obstacle on the primary slip plane.

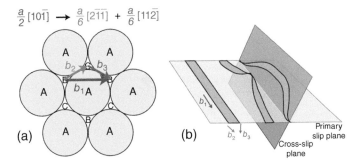

Fig. 4.17 (a) Plan view of a (111) plane in an fcc structure, showing how a dislocation with a perfect Burgers vector can dissociate into two partials and (b) depiction of how, in order to undergo cross-slip, such a dissociated pair of partials needs to be locally constricted into a single (pure screw) dislocation. Once it has passed onto the cross-slip plane, it can again dissociate into two partials (different from the pair on the primary slip plane).

There are some subtleties involved in cross-slip. One of them is that, in fcc structures, the "normal" **perfect dislocation**, with a Burgers vector of type $(a/2)$ <110>, is usually "**dissociated**" into two **partial dislocations**. This occurs because such a dissociation, which is shown in Fig. 4.17(a), is energetically favorable (according to Frank's rule – see Eqn. (4.8)). There is (elastic) repulsion between the two partials, but also attraction, since there is a **stacking fault** between them (with associated energy). This arises because, when one partial has passed, but not the other, the normal ABCABC stacking sequence in fcc now incorporates a short ACAC sequence, representing the hcp (hexagonal close-packed) structure. The equilibrium separation between partials is of the order of a few nm, but is greater when the **stacking fault energy,** γ, of the metal (difference between the energies of the fcc and hcp structures) is low. This is relevant to cross-slip because it can only occur if the two partials are (temporarily) forced back together to form a perfect (screw) dislocation – termed a **constriction**. This is illustrated in Fig. 4.17(b). This is easier in materials with high stacking fault energies. For example, Al, with a γ value of ~0.17 J m^{-2}, commonly undergoes cross-slip, whereas it is rare in Cu (γ ~0.05 J m^{-2}). This area has received extensive study over a long period [11–13].

Dislocations readily glide on slip planes, and perhaps on cross-slip planes, but edge dislocations can also move via a different mechanism, involving **vacancy migration** to the core (bottom row of extra half-plane), termed **climb**. In a similar way to the cross-slip of screw dislocations, it should be noted that all dislocations can readily change their edge/screw character, so climb is in effect a process available to all dislocations. This vacancy absorption causes the dislocation, or at least that segment of it where the vacancy is absorbed, to rise ("climb"), as shown in Fig. 4.18. Equally, the dislocation can move downwards by **emitting** a vacancy – also referred to as climb. This gives the dislocation much more freedom of motion than glide on its slip plane, but requires a lot of vacancy movement and tends to be relatively slow. However, this depends strongly on temperature, which dictates both the vacancy concentration and the

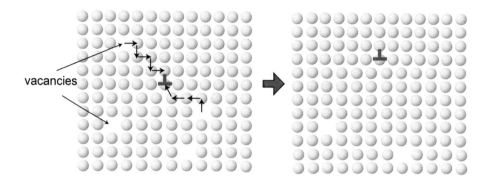

Fig. 4.18 Climb by absorption of vacancies at the core of an edge dislocation (in a simple cubic lattice).

frequency with which a vacancy moves (i.e. a neighboring atom jumps into it). It should also be understood that vacancies are much more mobile than atoms. At relatively high (homologous) temperatures, the mobility of vacancies is extremely high and this can allow dislocations to move more quickly, and with much greater freedom in terms of type of motion, than at lower temperatures. This is a prime reason why metals tend to be softer, and also to exhibit lower **work hardening rates**, at higher temperatures. This is, of course, widely exploited in **metal forming processes**, such as **rolling** and **forging**. In contrast to this, cross-slip does not have strong **thermal activation** and can occur at relatively low temperature (but does not confer the same freedom as climb in terms of type of motion).

4.2.4 Rearrangement of Dislocations and Recrystallization

Dislocations thus have freedom to move in different ways, driven by applied stresses or those from interactions between individual dislocations or between dislocations and other defects (grain boundaries, precipitates etc.). They move so as to reduce the energy – e.g. to minimize total stored elastic strain energy. This can lead to low energy configurations, with dislocations forming aligned arrays, by local rearrangements (via climb and glide). Some **annihilation** of dislocations may also occur. This is illustrated in Fig. 4.19 for a set of edge dislocations. Similar processes occur with screw and mixed dislocations (and, as emphasized above, any given dislocation can readily change its edge/screw proportions). These aligned dislocation arrays are called **polygon walls**, or **sub-grain boundaries**, and the process of forming them is called **polygonization** or **recovery** (because the associated drop in ρ leads to "recovery" of certain properties back to values typical of the material before ρ became high). A micrograph of a polyganized structure is shown in Fig. 4.19(b).

The reason for describing these **polygons** as **sub-grains** is that an organized array of dislocations of this type creates a (small) change between the orientation of the crystal planes on either side of the boundary. A set of edge dislocations of the type shown in Fig. 4.19 creates a **tilt boundary**, with the tilt angle given by

4.2 Further Dislocation Characteristics

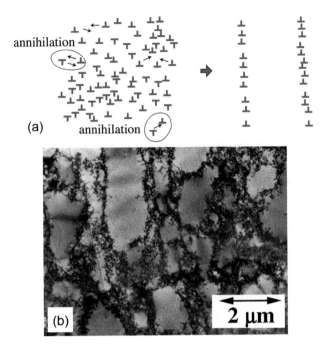

Fig. 4.19 (a) Schematic representation of the rearrangement (by glide and climb) of a set of parallel edge dislocations into a lower energy configuration (polygonization) and (b) TEM micrograph of a Cu sample after polygonization.

$$\sin\theta \sim \theta = \frac{b}{L} \tag{4.13}$$

where L is the (vertical) spacing between neighboring dislocations in the wall. These misorientation angles between sub-grains thus depend on the dislocation density before the reorganization occurred (and also on the sub-grain size), but they are typically of the order of a few degrees (with the sub-grain size typically being of the order of a micron). While a set of edge dislocations creates a tilt boundary, a similar set of screw dislocations creates a *twist boundary*. In practice, sub-grain boundaries commonly have both tilt and twist elements to the associated misorientation.

This process can in some cases progress a stage further, in the form of *recrystallization*. This term describes the phenomenon of the formation of a completely "new" set of grains (crystals). It occurs when one or more sub-grain boundaries attain a *misorientation* with their neighbors such that they are effectively conventional (high angle) grain boundaries, which usually happens when the angle reaches about 10° or so. Their structure is then no longer that of an array of dislocations and they are able to sweep through the surrounding region, eliminating the dislocation arrays as they go and creating new (more or less!) dislocation-free grains. This process is favored by high temperature (which greatly increases dislocation mobility) and also by a high original dislocation density – i.e. a high degree of *prior "cold work."* It is usually accompanied by a sharp drop in the hardness – much more dramatic than that associated with recovery (polygonization). Under some circumstances, it can occur dynamically during hot working, although more commonly it requires an extended heat treatment of some sort. There are several excellent reviews [14–16] covering various aspects of (dynamic) recrystallization.

4.3 Dislocations in Real Materials and Effects of Microstructure

4.3.1 Plastic Deformation of Polycrystals

The vast majority of metallic specimens or components are not in the form of single crystals, but are polycrystalline. Under an applied load, the Schmid factor differs for each grain. For randomly oriented grains (i.e. an **untextured** material), the average value of the Schmid factor is ~1/3 (often called the **Taylor factor**). It might thus be expected that σ_Y would have a value of about $3\tau_c$ (compared with about $2\tau_c$ for a single crystal). However, in practice the difference between the stress–strain behavior of single crystals and corresponding polycrystals is usually much greater than this. In a polycrystal, the deformation of each individual grain has to be compatible with that of its neighbors – i.e. there is a strong **constraint** effect. **Multiple slip** (the operation of more than one slip system) is normally required from the outset in virtually all grains, in order to satisfy this requirement, and substantially higher stresses are needed for yielding and plasticity, compared with single crystals. Also, the stress–strain curve shows no stages like those observed initially with a single crystal. Furthermore, the **yield point**, when plastic deformation becomes established throughout the sample, is in many cases not very sharply defined. This difference between the behavior of single crystals and polycrystals (under uniaxial load) is illustrated schematically in Fig. 4.20(a). The grain size is expected to have an effect, as is the presence of **crystallographic texture** (non-random distribution of the orientation of grains).

The bulk properties of the material will only be reflected in a test if the tested volume is large enough to capture these sensitivities. In particular, for polycrystals, only a relatively large assembly of grains is likely to satisfy this requirement. If, for example, the section of a small sample contains only single grains, or a handful of grains, then its response is likely to differ significantly from that of the bulk. There are several ways of highlighting this issue. One is shown [17] in Fig. 4.20(b), which presents experimental information obtained on (relatively coarse-grained) Al samples after two different degrees of compressive plastic deformation. These images are maps of equivalent plastic strain, obtained using the digital image correlation (DIC) technique. They highlight the inhomogeneous nature of the deformation, with some grains deforming much more than others. There are also marked variations within individual grains. Moreover, it can be seen that substantial changes take place in the morphologies of the grain boundaries: their local structure is likely to influence the way in which this occurs. It's clear that the overall (plastic) response of the material can only be well captured if an assembly of grains (with a representative texture, set of grain

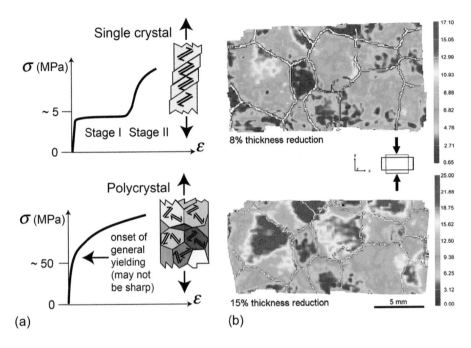

Fig. 4.20 Illustrations of the nature of plastic deformation in single crystals and polycrystals, showing (a) schematic representations of stress–strain curves and slip systems and (b) experimental DIC maps [17] giving spatial distributions of the (von Mises) equivalent plastic strain, obtained for a coarse-grained (columnar) Al polycrystal, after two different degrees of plastic deformation, as indicated.

boundaries etc.) is being deformed. This point is an important one, which is also emphasized elsewhere in the book, such as in §5.1.3.

The relationship between the internal mechanisms creating the plasticity and the observed stress–strain relationship, coupled via the microstructure of the metal, is thus a complex one. In fact, it is considerably more complex than the picture presented so far, since there are also likely to be various microstructural effects beyond that of the grain structure. These include the presence of solute atoms, phases other than that of the base metal, residual stresses etc. Some further details of these effects are presented in the sections below. Nevertheless, as outlined in §3.2.1, experimental stress–strain curves can often be captured quite effectively using empirical constitutive laws, despite the complexity of the mechanisms responsible for them.

4.3.2 Solution Strengthening (Substitutional Atoms)

The *"lattice friction stress"* (**Peierls stress**) is raised by the presence of solute atoms, which inhibit smooth gliding of dislocations. The effect on yield stress of substitutional solute is illustrated in Fig. 4.21, for Cu-Ni alloys, a system that exhibits complete solid solubility, so there are no complications from precipitate formation.

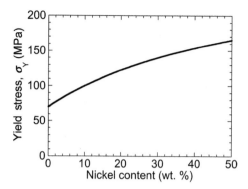

Fig. 4.21 Effect of Ni content on the yield stress of Cu-Ni alloys.

The strengthening effect is significant, but the changes are gradual and moderate in magnitude. This is a consequence of misfit strains being relatively small for substitutional solutes, and also being **purely hydrostatic** (no shear), so that interaction occurs only with edge dislocations (or edge components of mixed dislocations).

4.3.3 Effects of Interstitial Solute

Interstitial atoms usually exert stronger effects than substitutional ones, because they generate lattice strains that are: (a) larger in magnitude and (b) often shape-changing ("deviatoric") as well as dilational, so they interact with all dislocations. This is illustrated by the example of an interstitial atom located in an octahedral interstice in the bcc (body-centered cubic) structure – see Fig. 4.22. Furthermore, since interstitials are more mobile than substitutionals, they diffuse more rapidly and can segregate more quickly to dislocation lines, forming what are often called **Cottrell atmospheres** (after Alan Cottrell, a pioneer in many areas of physical metallurgy). With $\rho \sim 10^{14}$ m^{-2} (a typical value), the dislocation spacing is about 0.1 μm (see Eqn. (4.12)), and most solute atoms would need to move no more than a few tens of nm to reach one.

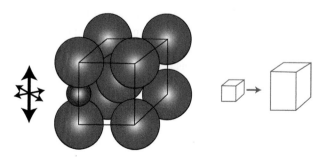

Fig. 4.22 An interstitial atom located in an octahedral interstice in a bcc structure, with a schematic depiction of how the associated distortion has a shear (shape-changing) component.

Interstitial solutes are not so common, although industrially important examples include carbon and nitrogen in steels and oxygen in copper. The case of oxygen in copper is interesting, since copper has a relatively high solubility for oxygen, leading to the possibility of creating a fine dispersion (§4.3.4) of (copper) oxide particles (via "*internal oxidation*") that is stable up to high temperatures [18, 19]. Moreover, even with no oxide formation, the presence of oxygen in solution affects the mechanical properties [20] of copper, and also its *recrystallization* (§4.2.4) behavior during hot working [21]. In general, a high level of dissolved oxygen impairs both the ductility and the conductivity, which are often important attributes for copper. It is thus common to refer to certain grades of commercially available copper as *oxygen-free high conductivity* (OFHC). In practice, the oxygen content of such material can vary over quite a wide range – perhaps from 2 ppm up to about 200 ppm. The properties of these materials can thus vary quite significantly, particularly in the work-hardened (as-rolled, as-extruded etc.) state.

Another well-known example of a system in which interstitials can play a significant role is that of mild steel, which is effectively pure iron (α-Fe) with just a small concentration of carbon present. (In fact, what is often termed mild steel usually contains other constituents, such as Si and Mn, with the result that the effect may be weak or absent: it is likely to be more noticeable in what would usually be regarded as high purity Fe.) Diffusion distances ($\sim\sqrt{(Dt)}$) of C atoms in mild steel are shown in Fig. 4.23. At room T, atmospheres should form in a few hours. At 150 °C, that time falls to a few seconds and at 300 °C formation would be virtually instantaneous.

Furthermore, saturated atmospheres can form with very low solute contents. For example, with $\rho \sim 10^{14}$ m^{-2}, a solute level of 0.001% is sufficient to provide an interstitial solute atom in every cross-sectional atom plane containing a dislocation. The formation of a carbon atmosphere is shown schematically in Fig. 4.24. Some examples of the effects that these atmospheres can have on the mechanical behavior are shown schematically in Fig. 4.25. Several characteristic features can be seen on such stress–strain plots. These can be understood in terms of the pinning effect exerted by the atmospheres and the ease with which they can form. (The "serrations" in Fig. 4.25(b) are due to repeated formation of atmospheres, and subsequent escape of

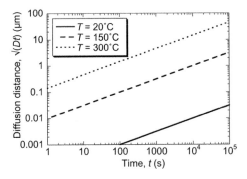

Fig. 4.23 Diffusion distances for (interstitial) carbon in ferrite (α-iron).

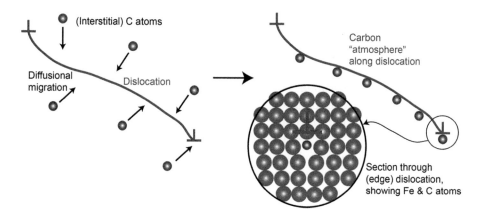

Fig. 4.24 Schematic depiction of the formation of a carbon atmosphere along the length of a dislocation in a steel.

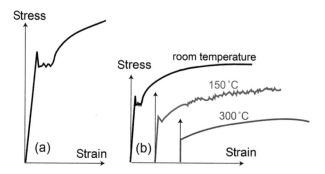

Fig. 4.25 Schematic stress–strain plots for a low C steel, showing effects of interstitial solute atmospheres around dislocations: (a) upper and lower yield points, observed at room temperature, and (b) the Portevin–Le Chatelier effect, observed on raising the temperature.

dislocations from them, during the test: this is termed the "***Portevin–Le Chatelier effect***.") These and other effects associated with Cottrell atmospheres are well covered in the literature [22–25]. Further details are provided in §5.3.4, including information about the so-called "***Lüders bands***" that sometimes form during the burst of plasticity that follows the escape of dislocations from their atmospheres.

It might be noted at this point that curves with shapes such as those in Fig. 4.25 (particularly concerning the initial load drop) cannot be accurately captured using either Ludwik–Hollomon or Voce constitutive laws (§3.2.1). Indeed, it should be recognized that there are other types of shape, such as those characteristic of single crystals (Fig. 4.11), for which this is also true. Other types of constitutive law could be used for these, but it may be noted that such shapes are not very common. Single crystals are certainly not encountered very often. Mild steel sounds as if it should be much more ubiquitous, but in fact, as outlined in §5.3.4, only rather few ferrous metals, including high purity iron ("Armco Iron"), have stress–strain curves with shapes of the type shown in Fig. 4.25. Moreover, even those often exhibit load drops and subsequent bursts of straining at constant load that are less well defined than those in the schematic plots of Fig. 4.25. The vast majority of steels, including many that might be termed mild steel, actually have at least slightly higher C contents, and also other constituents such as Si and Mn, that have the effect of removing this "upper and lower yield point" effect (although the concept of dislocations being "pinned" by interstitial atoms remains valid).

4.3.4 Precipitation Hardening and Dispersion Strengthening

As outlined above, dislocations can be pinned, or their mobility reduced, by the presence of solute atoms. However, it's also possible for them to be impeded, often more effectively, by slightly larger scale obstacles. These include (sessile) jogs, dislocation tangles etc., grain boundaries and second phases (e.g. in the form of precipitates). A common obstacle in metallic alloys is a precipitate or inclusion, which will normally have a different crystal structure from that of the matrix, so that there is no possibility of the dislocation passing through it. A dislocation can, however, **bow** around an obstacle, leaving a **dislocation loop** around it (via a similar mechanism to that involved in operation of a Frank–Read source) – see Fig. 4.26. This is often termed **Orowan bowing** (after Egon Orowan).

By-passing of obstacles by gliding in the slip plane (Fig. 4.26) may require the dislocation to adopt sharp **curvature**, needing additional energy and hence a higher applied stress. The peak in curvature occurs when the dislocation forms a semi-circle, with two obstacles on a diameter – see Fig. 4.27. For a segment of line dl, the force acting normal to the segment is F dl, so the net force in the vertical direction is thus

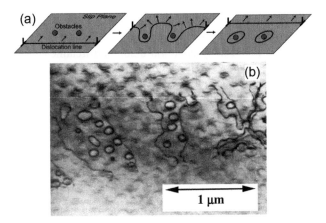

Fig. 4.26 (a) Schematic representation of Orowan bowing of a dislocation around a pair of obstacles on the slip plane, leaving dislocation loops around them, and (b) a TEM micrograph showing such loops in a "Waspalloy" Ni-based superalloy.

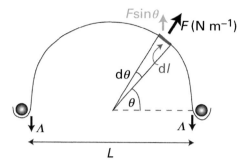

Fig. 4.27 Force on a short segment of a dislocation, at peak curvature, bowing past two obstacles.

$$\int F \sin\theta \, dl = \int_0^\pi F \sin\theta \left(\frac{L}{2} d\theta\right) = \frac{FL}{2}[-\cos\theta]_0^\pi = FL = \tau bL \quad (4.14)$$

This force is balanced by the **line tension**, Λ, acting along the dislocation line at both ends of the semi-circle. This line tension (in N) is equivalent to the energy per unit length (in J m^{-1}) of the dislocation, given by $G b^2/2$ (see Eqn. (4.7)). It follows that

$$\tau bL = 2\frac{1}{2}Gb^2 \therefore \tau = \frac{Gb}{L} \quad (4.15)$$

This is termed the **bowing stress**, or the **Orowan stress**. Equation (4.15) gives the shear stress needed to cause dislocations to become sufficiently curved to avoid obstacles on the slip plane spaced L apart. Since $G \sim 30$ GPa and $b \sim 0.3$ nm, an obstacle spacing of 10 µm corresponds to a small increment of stress (~1 MPa), whereas a spacing of 0.1 µm requires a relatively large stress (100 MPa).

The most common method of producing a suitably fine set of obstacles is to promote the formation of precipitates under conditions such that they are very small and closely spaced. This is normally done via a solution treatment, quenching and ageing sequence. Much of the pioneering work was done on the Al-Cu system, which still forms the basis of commercially important precipitation-hardened alloys. The phase diagram is shown in Fig. 4.28. Typically, a three-stage process is employed, consisting of (a) solution treatment at T_{ST} (producing single phase α), (b) quenching to room (ambient) temperature, T_{amb} (giving a supersaturated solid solution) and (c) holding at T_{age} (ageing temperature), to allow sufficient diffusion to form (fine scale) precipitates.

In practice, the expected θ (Al$_2$Cu) phase (Fig. 4.28) is often not the first phase to form during heat treatment. It is common for the phases that form initially to be less stable thermodynamically than θ, but to be nucleated more readily – usually because they are closer in structure to the matrix and hence form relatively low energy interfaces with it. In fact, the structures initially formed, particularly at relatively low ageing T, are sometimes termed "zones" (**Guinier–Preston zones**), since they are really just regions of the matrix with different compositions, rather than distinct phases. These interfaces may be fully coherent, giving very low interfacial energies. Other metastable precipitates (often termed θ′ and θ″) may then form from these zones, having semi-coherent interfaces, before they finally transform to the equilibrium (θ) phase, which has an incoherent interface with the matrix. Simultaneously with these changes, *coarsening* often occurs, so the spacing between precipitates increases. Some illustrative micrographs are shown in Fig. 4.29. There has been extensive work [26–28] on the details of these effects.

There are several ways in which the presence of these precipitates can affect the motion of dislocations, and hence the strength (yield stress and work hardening rate). For example, when the interface is coherent or semi-coherent, the associated *coherency strains* in the adjacent lattice inhibit dislocation glide. The zones and precipitates that form initially are often small and closely spaced (so τ would be high if they were

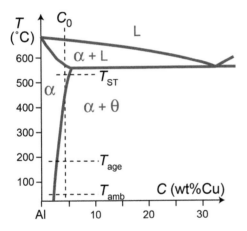

Fig. 4.28 The Al-Cu phase diagram, with the composition of a typical age hardening alloy marked as C_0 and heat treatment temperatures also indicated.

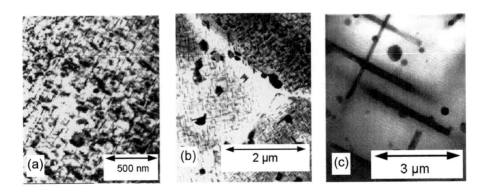

Fig. 4.29 TEM micrographs of an Al-4wt%Cu alloy after increasing heat treatment times, showing (a) GP zones, (b) GP zones and θ′ precipitates and (c) θ′ and θ precipitates.

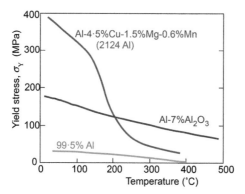

Fig. 4.30 Representative dependence of yield stress on temperature, for commercial purity Al and for precipitation-hardened (2124) and dispersion-strengthened Al alloys.

effective barriers), but dislocations may be able to **cut through** them, particularly if they have a similar crystal structure to that of the matrix. For precipitates forming later in the sequence, this is not possible and dislocations are forced to bow around them. This becomes easier as the precipitate array becomes coarser ("**over-ageing**"). The combination of these effects can lead to relatively complex changes in hardness during ageing.

Precipitation hardening is commonly used to generate strength at room T, but is unable to provide **high T strength**. Solution treatment and ageing is a convenient way of creating a fine precipitate array, but it requires that the precipitates **dissolve** at elevated T. Even at moderate temperatures, which cause little dissolution, precipitates tend to **coarsen** sufficiently to make age hardening ineffective.

It is possible to generate particles that are stable at high T, but they are usually not as fine or closely spaced as in optimally age-hardened alloys. Commonly, these are **oxide particles**, which may be created by internal oxidation of the matrix or by mixing of matrix and oxide powders. This is often termed **dispersion strengthening**. There are several ways in which relatively fine dispersions of such insoluble particles can be created, often involving the processing of powders. For example, aluminum powder particles, with surface layers of alumina (perhaps thickened by heating in air), can be consolidated by extrusion. This breaks the oxide layers up into fine platelets (probably aligned parallel to the extrusion direction). The plots in Fig. 4.30 show typical variations of strength with T for different types of Al alloy.

4.4 Deformation Twinning and Martensitic Phase Transformations

Dislocation glide (perhaps assisted by cross-slip and/or climb) is not the only mechanism of plastic deformation (although it is the predominant one, particularly for metals at room T). For example, dislocations cannot exist in amorphous materials, which can nevertheless deform plastically – usually by formation of **shear bands** on planes where the shear stress is a maximum (at 45° to the loading axis for uniaxial loading). Some metals can exist with an amorphous atomic scale structure (which resembles that of a liquid), although they are far from common. In any event, while sliding on shear bands is in some ways similar to dislocation glide, the mechanism does not operate easily and most amorphous materials do not undergo plastic deformation very readily (and hence tend to be brittle). Plastic deformation can also take place via diffusional rearrangement of atoms, with or without dislocations being involved, so that the deformation is time-dependent (and T-dependent). Such "creep" behavior is covered below in §4.5.

4.4.1 Deformation Twinning

There is also a completely different kind of deformation that can occur in crystalline materials, over a range of T. This involves the **homogeneous shear** of large volumes of material. It is a **cooperative** process, such that each atom moves a small, well-defined distance relative to its neighbors. This shearing may create a new crystal structure (phase), in which case the process is termed a **martensitic** phase transformation – see §4.4.2. In other cases, the shearing re-creates the parent phase in a different orientation – this is termed a **twin**. Such twins are called **deformation twins** or **mechanical twins**. The various types of deformation mechanism are depicted schematically in Fig. 4.31.

Figure 4.32 compares plastic deformation generated by slip (dislocation glide) with that resulting from the shearing process associated with deformation twinning. Slip leaves the lattice unaffected and relatively large shear displacements can be generated on a single slip plane, if many dislocations pass across it. (This commonly happens, creating **persistent slip bands**, which are sometimes visible on free surfaces – see, for example, Fig. 4.10, Fig. 7.12 and Fig. 9.13.) Twinning can generate a similar type of displacement, but it is usually of relatively small magnitude and it is homogeneously distributed through the twinned region. Deformation twinning has associated **twinning elements**, analogous to the slip system for dislocation glide. These are the **twinning plane** and the **twinning direction**. Atoms move in the twinning direction by a prescribed distance, generating the **twinning shear** (ratio of distance moved by one twinning plane relative to its neighbor, divided by the inter-planar spacing).

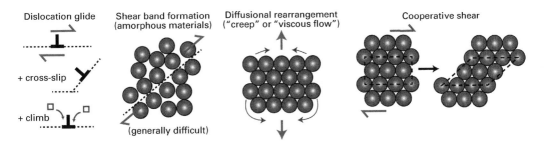

Fig. 4.31 Schematic representation of different mechanisms of plastic deformation.

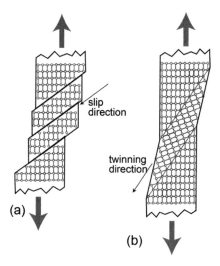

Fig. 4.32 Plastic deformation by (a) slip and (b) twinning, showing the associated changes in lattice orientation.

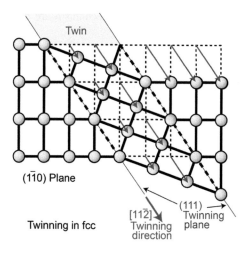

Fig. 4.33 Crystallography of deformation twinning in fcc.

The twinning elements for fcc are shown in Fig. 4.33. It can be seen that the structure in the twin is a mirror image (across the twin plane) of that in the parent grain. It is also apparent that the interface between parent and twin in this arrangement is a *fully coherent boundary* (i.e. every atom is in its correct location with regard to both parent and twin lattices). The energy of such a coherent boundary is very low – hence the tendency for twin boundaries to be straight and parallel, since their energy is minimized when they are in this orientation.

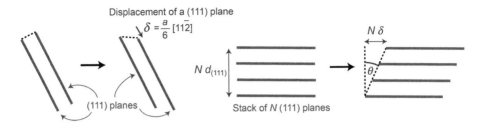

Fig. 4.34 Geometry of the twinning shear in fcc.

The twinning shear for a particular case can be obtained by considering the displacements induced as twinning occurs. This is illustrated in Fig. 4.34 for the fcc case. It can be seen that the twinning shear is given by

$$\text{twinning shear } (= \tan\theta) = \frac{N\delta}{Nd_{(111)}} = \frac{\delta}{d_{(111)}}$$

$$|\delta| = \frac{a\sqrt{6}}{6} = \frac{a}{\sqrt{6}}, \quad d_{(111)} = \frac{a}{\sqrt{3}} \quad (4.16)$$

$$\therefore \text{ twinning shear } = \frac{1}{\sqrt{2}}$$

Deformation twinning is common in structures of relatively low crystallographic symmetry. They tend to have fewer independent slip systems, making it less likely that an imposed strain can be fully accommodated by dislocation glide. For example, **hexagonal metals** are more prone to deformation twinning than cubic metals. Similarly, deformation of tin (**tetragonal**) occurs predominantly by twinning, causing the so-called "**tin cry**" (acoustic waves generated by twin nucleation and growth) – an effect that has been known [29] for many decades.

The shear stress needed to trigger deformation twinning varies, but it is usually higher than that for dislocation glide. However, since twinning involves cooperative, simultaneous motion of atoms by small distances relative to neighbors, it occurs very quickly. In fact, atoms move at ~ the **speed of sound**, so twin formation is effectively instantaneous. In contrast, dislocation glide, which is not a cooperative process, cannot readily generate large plastic strains very quickly. High imposed strain rates thus favor deformation twinning. For example, deformation twins (sometimes called **Neumann bands**) may be observed [30] in steel (ferrite) subjected to very high strain rates (e.g. in explosive or impact events). Since they do not normally occur at ambient T in such bcc structures, their existence may be used as forensic evidence about deformation conditions in crashes, explosions etc.

Thermal activation is not needed for deformation twinning, whereas dislocation glide does require some thermal energy, particularly if climb is necessary. Deformation twinning thus becomes more likely as the temperature falls, and at very low T it can predominate in materials for which it is normally rare, such as cubic metals. This is illustrated schematically in Fig. 4.35. In general, deformation twinning is favored by anything that inhibits dislocation mobility, including the presence of efficient pins, barriers etc.

Twins can also form without mechanical deformation, particularly at high T. These are usually termed **annealing twins** or **growth twins**. They are distinguishable from deformation twins via their morphologies. Deformation twins tend to be lenticular (lens-shaped), since they are constrained at their ends (grain boundaries). If they were to remain parallel-sided, then the shear associated with their formation would lead to voids at the grain boundaries. Annealing twins form by diffusional, reconstructive processes, with no shear involved. They thus tend to be entirely parallel-sided, with fully coherent boundaries throughout. The micrographs in Fig. 4.36 illustrate these differences.

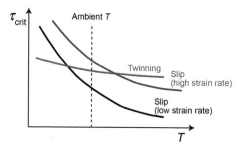

Fig. 4.35 Schematic representation of the dependence on temperature and strain rate of the critical shear stress necessary to stimulate slip or twinning.

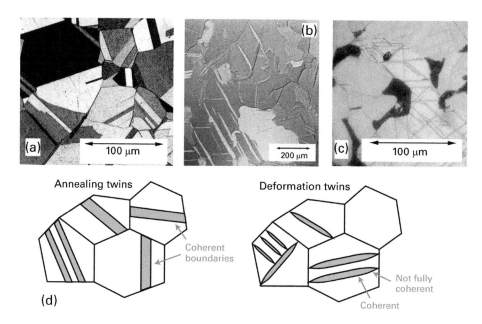

Fig. 4.36 Optical micrographs of (a) annealing twins in Cu-30wt%Zn (brass), (b) deformation twins in cold-rolled Zn (hexagonal) and (c) Neumann bands in a 0.2wt%C steel subjected to very high strain rate (in an explosion). Also shown is (d) a schematic representation of the different morphologies typical of the two types of twinning.

4.4.2 Martensitic Transformations

Shear of the type induced during deformation twinning can also create a crystal structure differing from that of the parent grain. As a simple example, if the shear depicted in Fig. 4.33 is imposed, not on every (111) plane, but only on every second (111) plane, then the structure is transformed from fcc to hcp – see Fig. 4.37. Other martensitic transformations are crystallographically more complex (including the important one occurring in the Fe-C system[1]). Martensitic transformations exhibit similarities to deformation twinning, and also differences.

Because martenstic transformations occur in such a way that *atomic correspondence* is retained (i.e. all of the atoms in the transformed region have the same neighbors as before), and since it may be possible to create a driving force for reversal of the phase change, the interesting possibility arises of *recovering* the *original shape*, in a way that is not possible with other mechanisms of plastic deformation. In practice, the characteristics required in order for this to be possible limit such effects to a relatively narrow range of alloys and conditions. Nevertheless, the phenomenon is now quite widely exploited in industrial applications.

When a Ni–50%Ti alloy is mechanically loaded in a certain T range, it can exhibit large, apparently plastic, strains, which spontaneously reverse on unloading – see

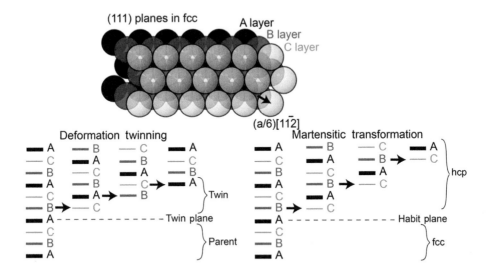

Fig. 4.37 Stacking sequence changes during shearing of (111) planes in the fcc structure to form either a twin or the hcp structure.

---

[1] The phase is named after Adolf Martens, who worked on steels in Berlin during the late nineteenth century. Formation of Fe-C martensite requires relatively rapid cooling and/or deformation, the phase is metastable and it has a high hardness. However, none of these features are characteristic of all martensitic phases, or even the majority of them. It's only relatively recently that martensitic transformations have become fully understood, and their characteristics directly exploited in applications such as shape memory devices.

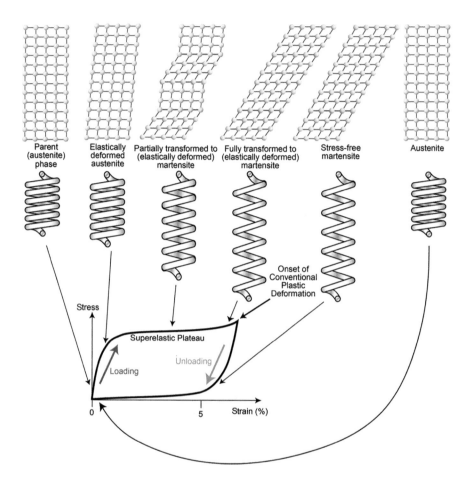

Fig. 4.38 Schematic representation of the stress–strain relationship, and associated changes in structure, for a material exhibiting superelastic deformation.

Fig. 4.38. This is often termed "*superelasticity*." These strains (several %) are generated by the lattice shearing to form a new ("martensite") phase. Increases in the transformed proportion can be stimulated without much increase in applied stress, giving a characteristic "superelastic plateau." An experimental stress–strain curve of this type is shown in §3.2.3 (Fig. 3.7(b)), where it can be seen that, in practice, the stress does tend to rise somewhat during this regime.

Superelastic (SE) deformation can only occur if the stress to form martensite is lower than that needed for conventional plastic deformation (dislocation glide). Both processes are affected by temperature. Commonly, the martensitic phase is more stable at lower T, while the austenitic phase is favored at higher T. Also, there is usually **hysteresis** in the transformation – i.e. both start and finish temperatures differ for heating and cooling – see Fig. 4.39(a) (and Fig. 3.7(a) in §3.2.3). For superelasticity to be possible, the temperature must be above A_f (so the austenitic phase is

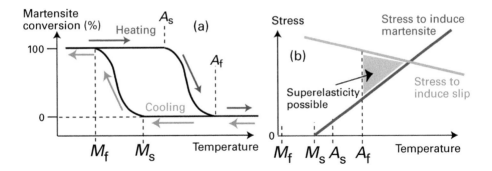

Fig. 4.39 Schematic representation of transformation characteristics in a system exhibiting superelasticity: (a) transformation temperatures (in the absence of stress) and (b) stress–temperature domain for superelasticity.

initially predominant), but low enough to ensure that the stress to form martensite is below that needed for plastic deformation (slip) – see Fig. 4.39(b). In practice, this temperature range is often rather limited.

The so-called "*shape memory effects*" (SMEs) also involve martensitic transformations, but these are stimulated, not by imposed strain, but by changes of temperature. They also involve a new concept – that of material being "*trained*" to have a preferred shape. This involves heating the specimen (while constrained into a certain shape) to high T (well above A_f), holding it at this T for a short period and then cooling it quickly to ambient T. Stress relaxation occurs during holding and then, during cooling (in the constrained shape), the austenite–martensite transformation takes place in such a way as to minimize the overall shape change. (When a portion of the lattice shears to form martensite, there are usually several alternative directions in which it can do this – forming different "variants" – so groups of variants can form which, taken together, have a similar shape to the original parent material.) There is subsequently a tendency for the specimen to adopt its "trained" shape, in which transformation between parent and martensitic phases takes place readily. Recovery of a "trained" shape by heating is illustrated in Fig. 4.40. There is extensive literature [31–33] on SME and SE alloys.

4.5 Time-Dependent Deformation (Creep)

4.5.1 Background

For all of the mechanisms described so far in this chapter, there is an underlying assumption that they take place instantaneously. This is implicit in the identification of "stress–strain" relationships, which incorporate no reference to time. However, as described in §3.2., there is also recognition that, in practice, there are some circumstances in which the plastic deformation of metals does in fact exhibit a time-dependence and constitutive laws are available to capture this behavior. One of these

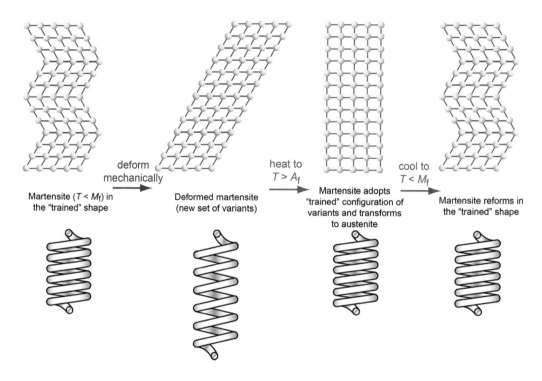

Fig. 4.40 SME recovery of a "trained" shape, after a large (apparently plastic) strain in the martensitic state, by heating the specimen, transforming the martensite to austenite.

scenarios relates to situations in which a sample is loaded very quickly – for example as a result of **ballistic impact** or **rapid forging**. It is generally observed that the stress needed to effect a given strain is raised under these circumstances – see §3.2.2. The mechanistic explanation for this is often that it is difficult for the necessary dislocation motion to take place quickly enough to create the strain concerned, at least without the driving force being raised somewhat. It's also possible for a switch in mechanism to occur, for example so that deformation twinning or martensitic phase changes (both of which can occur more quickly than dislocation glide) start to make a significant contribution, as outlined in §4.4.1 and §4.4.2.

However, there is another type of dependence on time that is also of considerable practical importance, which usually relates to much longer time scales. Depending on a number of factors, it is sometimes observed that progressive plastic deformation can take place if a sample or component is subjected to stress over an extended period, even if the stress is below the yield stress (i.e. in a regime where, in principle, only elastic deformation should occur). Such deformation is commonly referred to as **creep**. The detailed mechanisms responsible for creep tend to be complex. Only a very superficial overview is given here and other sources [34–39] should be consulted for more details. However, they almost always involve diffusion of some sort. This is how the time-dependence arises, since diffusional processes are progressive with time.

4.5 Time-Dependent Deformation (Creep)

Empirical equations (constitutive laws) for creep are presented in §3.2.4, where it is noted that the initial ("primary") creep rate is often considerably higher than the steady state rate that may be established later. These "laws," and the tendency for the rate to fall off with time (under constant applied true stress), tend to be broadly valid for all creep processes, although it is sometimes observed that the range of values found to be appropriate for the stress exponent, n, and the activation energy, Q, is characteristic of the different types of creep mechanism.

4.5.2 Coble Creep

Creep deformation may or may not involve various defects, particularly ***dislocations*** or ***grain boundaries***. These might simply act as ***fast diffusion paths***, depending on factors such as dislocation density, grain size/shape and temperature. The shape change experienced by the sample may arise simply from atoms (molecules) becoming redistributed by diffusion. When this occurs on the scale of a grain, with the diffusion occurring mainly via grain boundaries, then this is commonly referred to as ***Coble creep***. It is illustrated schematically in Fig. 4.41.

Raising the applied stress accelerates the rate of diffusion and hence the creep rate. The driving force for this net migration of material (from the "***equatorial***" regions of grains to the "***polar***" regions) is that an applied tensile stress like this creates hydrostatic compression in the equatorial regions and hydrostatic tension in the polar regions. The atoms then tend to move from the more "crowded" to the more "open" regions. The diffusive flux can be considered as migration of vacancies (in the opposite direction to that of atomic motion), although the concept of vacant sites is less well-defined in a grain boundary than in the lattice.

Raising the temperature also increases the creep rate. This is simply due to the rates of diffusion becoming higher as a consequence of the ***Arrhenius*** dependence. It may

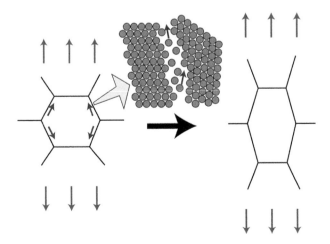

Fig. 4.41 Schematic depiction of the mechanism of Coble creep.

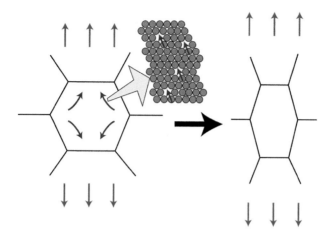

Fig. 4.42 Schematic depiction of the mechanism of Nabarro–Herring creep.

be noted, however, that, since the activation energy for diffusion through grain boundaries is low (and most of the diffusion takes place via such paths), this type of creep is often the dominant one at *relatively low temperatures*. It is also favored by a *fine grain size*.

4.5.3 Nabarro–Herring Creep

A similar type of creep deformation to that described above can occur with the diffusion being predominantly within the interior (crystal lattice) of the grains, rather than in the grain boundaries. This is often termed ***Nabarro–Herring creep***. It is depicted in Fig. 4.42. There is a considerably greater sectional area available via crystal lattice paths, particularly if the grains are relatively large. On the other hand, the activation energy is higher, so diffusion rates tend to be low, particularly at low temperature. This type of creep thus tends to dominate over Coble creep at *relatively high temperature* (and with *large grains or single crystals*).

4.5.4 Dislocation Creep

Purely diffusional creep (Coble and Nabarro–Herring) is fairly simple, and does occur under certain conditions – usually with relatively low levels of applied stress. With higher stresses (relative to the yield stress at the temperature concerned), it is common for a type of creep to occur that involves motion of dislocations (particularly in metals, where dislocation densities tend to be high). Provided the stress is below the yield stress, conventional macroscopic plasticity, occurring predominantly via dislocation glide, should not occur. However, with stresses that are starting to approach the yield stress, and are maintained for extended periods, progressive dislocation motion, and hence macroscopic plastic deformation, can occur, often facilitated by extensive climb

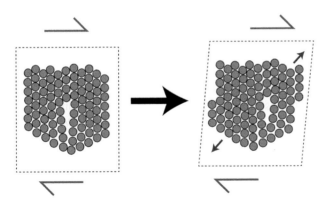

Fig. 4.43 Schematic depiction of the mechanism of dislocation creep.

(absorption or emission of vacancies at the core) of individual dislocations. This is shown schematically in Fig. 4.43.

In detail, there are several different ways in which combinations of dislocation glide and diffusion in the vicinity of dislocations can promote progressive (time-dependent) deformation (creep). Some of these have been given specific names, but these often relate to observed dependences on the main variables (***temperature***, ***stress*** and ***grain size***), rather than being clear about the precise mechanisms involved. In general, they all involve some combination of dislocation climb and glide, although, in particular cases, factors such as the presence of **obstacles** (e.g. fine **precipitates** or **phase boundaries**), the ease of **cross-slip** etc. may affect the observed behavior. The initial ***dislocation density*** may also be relevant, although this could change during the process. In fact, it should always be borne in mind that ***microstructural change*** might occur ***during creep*** (as a result of extended periods at high temperature), affecting the mechanical response. For example, ***recrystallization*** (§4.2.4) often changes the microstructure dramatically, sharply reducing the dislocation density. It might be imagined that this would reduce the rate of dislocation creep, but in fact it might well increase, since there will probably be a sharp reduction in the yield stress, promoting the onset of conventional plasticity (extensive dislocation glide).

4.5.5 Effects of Microstructure on Creep

In view of the complexity of creep mechanisms, it is difficult, if not impossible, to predict creep characteristics from a knowledge of the microstructure. In fact, the same could be said of plasticity (yield stress and work hardening characteristics). As with plasticity, however, ***guidelines*** can be identified concerning features likely to affect (inhibit) creep, and some of these are similar for the two. For example, a ***fine array of precipitates***, which will inhibit dislocation glide and hence raise the yield stress, is also likely to inhibit (dislocation) creep. However, there are limits to such linkage. For example, precipitates might ***dissolve*** at the high temperatures involved in creep. More fundamentally, some features can affect creep and plasticity quite differently. For example, while a fine grain size tends to raise the yield stress, as a result of **grain boundaries** acting as **obstacles** to dislocation glide, it can cause accelerated creep in a diffusion-dominated regime, since such boundaries also constitute **fast diffusion paths**.

In any event, as with plasticity, (empirical) **constitutive laws** are used to model and predict creep outcomes (§3.2.4). There is sometimes scope for interpreting the **values of parameters** in these laws in terms of the dominant **mechanisms** involved. In particular, if rates of creep are measured over a range of temperature, then it may be possible to evaluate the **activation energy**, Q (in an Arrhenius expression), which could in turn provide information about the type of diffusional process that is rate-determining. It is also often claimed that the value of a **stress exponent**, n, obtained via creep rate measurements over a range of applied stress, is indicative of the dominant mechanism, with a low value (~1–2) indicating **pure diffusional creep** and a higher value (~3–6) suggestive of **dislocation creep**. The theoretical basis for such conclusions may sometimes be questioned, and in any event these laws must be recognized as essentially empirical, but it is certainly important to be able to characterize the creep response, and to be clear about the regime of temperature and stress for which a particular law is valid.

References

1. Hirth, JP, A brief history of dislocation theory. *Metallurgical Transactions A: Physical Metallurgy and Materials Science*, 1985. **16**(12): 2085–2090.
2. Rodriguez, P, Sixty years of dislocations. *Bulletin of Materials Science*, 1996. **19**(6): 857–872.
3. Chang, R and LJ Graham, Edge dislocation core structure and Peierls barrier in body-centred cubic iron. *Physica Status Solidi*, 1966. **18**(1): 99–103.
4. Schmid, E and W Boas, Plasticity of Crystals, with Special Reference to Metals *(translated from German)*. London: FA Hughes & Co., 1950.
5. Tang, M, B Devincre and LP Kubin, Simulation and modelling of forest hardening in body centre cubic crystals at low temperature. *Modelling and Simulation in Materials Science and Engineering*, 1999. **7**(5): 893–908.
6. Saimoto, S, Dynamic dislocation-defect analysis. *Philosophical Magazine*, 2006. **86**(27): 4213–4233.
7. Norfleet, DM, DM Dimiduk, SJ Polasik, MD Uchic and MJ Mills, Dislocation structures and their relationship to strength in deformed nickel microcrystals. *Acta Materialia*, 2008. **56**(13): 2988–3001.
8. Brown, LM, Constant intermittent flow of dislocations: central problems in plasticity. *Materials Science and Technology*, 2012. **28**(11): 1209–1232.
9. Ashby, MF, Results and consequences of a recalculation of Frank–Read and Orowan stress. *Acta Metallurgica*, 1966. **14**(5): 679–681.
10. Xu, SZ, LM Xiong, YP Chen and DL McDowell, An analysis of key characteristics of the Frank–Read source process in FCC metals. *Journal of the Mechanics and Physics of Solids*, 2016. **96**: 460–476.

11. Thornton, PR, TE Mitchell and PB Hirsch, Dependence of cross-slip on stacking fault energy in face centred cubic metals and alloys. *Philosophical Magazine*, 1962. **7**(80): 1349–1369.
12. Puschl, W, Models for dislocation cross-slip in close-packed crystal structures: a critical review. *Progress in Materials Science*, 2002. **47**(4): 415–461.
13. Rao, SI, DM Dimiduk, TA Parthasarathy, J El-Awady, C Woodward and MD Uchic, Calculations of intersection cross-slip activation energies in FCC metals using nudged elastic band method. *Acta Materialia*, 2011. **59**(19): 7135–7144.
14. Sakai, T and JJ Jonas, Dynamic recrystallization – mechanical and microstructural considerations. *Acta Metallurgica*, 1984. **32**(2): 189–209.
15. Doherty, RD, DA Hughes, FJ Humphreys, JJ Jonas, DJ Jensen, ME Kassner, WE King, TR McNelley, HJ McQueen and AD Rollett, Current issues in recrystallization: a review. *Materials Science and Engineering A: Structural Materials Properties Microstructure and Processing*, 1997. **238**(2): 219–274.
16. Sakai, T, A Belyakov, R Kaibyshev, H Miura and JJ Jonas, Dynamic and post-dynamic recrystallization under hot, cold and severe plastic deformation conditions. *Progress in Materials Science*, 2014. **60**: 130–207.
17. Roters, F, P Eisenlohr, L Hantcherli, DD Tjahjanto, TR Bieler and D Raabe, Overview of constitutive laws, kinematics, homogenization and multiscale methods in crystal plasticity finite-element modeling: theory, experiments, applications. *Acta Materialia*, 2010. **58**(4): 1152–1211.
18. Lee, J, YC Kim, S Lee, S Ahn and NJ Kim, Correlation of the microstructure and mechanical properties of oxide-dispersion-strengthened coppers fabricated by internal oxidation. *Metallurgical and Materials Transactions A: Physical Metallurgy and Materials Science*, 2004. **35A**(2): 493–502.
19. Chen, F, ZQ Yan and T Wang, Effects of internal oxidation methods on microstructures and properties of Al_2O_3 dispersion-strengthened copper alloys, in *High Performance Structural Materials*, Han, Y, ed. Singapore: Springer-Verlag, 2018, pp. 1–8.
20. Guskovic, D, D Markovic and S Nestorovic, Effect of deformation and oxygen content on mechanical properties of different copper wires. *Bulletin of Materials Science*, 1997. **20**(5): 693–697.
21. Ravichandran, N and Y Prasad, Influence of oxygen on dynamic recrystallization during hot working of polycrystalline copper. *Materials Science and Engineering A: Structural Materials Properties Microstructure and Processing*, 1992. **156**(2): 195–204.
22. Takeuchi, S and AS Argon, Glide and climb resistance to the motion of an edge dislocation due to dragging a Cottrell atmosphere. *Philosophical Magazine A: Physics of Condensed Matter Structure Defects and Mechanical Properties*, 1979. **40**(1): 65–75.
23. Zhao, JZ, AK De and BC De Cooman, Kinetics of Cottrell atmosphere formation during strain aging of ultra-low carbon steels. *Materials Letters*, 2000. **44**(6): 374–378.
24. Wilde, J, A Cerezo and GDW Smith, Three-dimensional atomic-scale mapping of a Cottrell atmosphere around a dislocation in iron. *Scripta Materialia*, 2000. **43**(1): 39–48.
25. Veiga, RGA, M Perez, CS Becquart and C Domain, Atomistic modeling of carbon Cottrell atmospheres in BCC iron. *Journal of Physics-Condensed Matter*, 2013. **25**(2).
26. Herman, H, ME Fine and JB Cohen, Formation and reversion of Guinier–Preston zones in Ai-5.3 at.% Zn. *Acta Metallurgica*, 1963. **11**(1): 43–56.
27. Gerold, V, On the structures of Guinier–Preston zones in Al-Cu alloys. *Scripta Metallurgica*, 1988. **22**(7): 927–932.

28. Matsuda, K, H Gamada, K Fujii, Y Uetani, T Sato, A Kamio and S Ikeno, High-resolution electron microscopy on the structure of Guinier–Preston zones in an Al-1.6 mass pct Mg_2Si alloy. *Metallurgical and Materials Transactions A: Physical Metallurgy and Materials Science*, 1998. **29**(4): 1161–1167.
29. Chalmers, B, The cry of tin. *Nature*, 1932. **129**: 650–651.
30. Kelly, A, Neumann bands in pure iron. *Proceedings of the Physical Society of London Section A*, 1953. **66**: 403–405.
31. Otsuka, K and X Ren, Physical metallurgy of Ti-Ni-based shape memory alloys. *Progress in Materials Science*, 2005. **50**(5): 511–678.
32. Kim, HY, Y Ikehara, JI Kim, H Hosoda and S Miyazaki, Martensitic transformation, shape memory effect and superelasticity of Ti-Nb binary alloys. *Acta Materialia*, 2006. **54**(9): 2419–2429.
33. Jani, JM, M Leary, A Subic and MA Gibson, A review of shape memory alloy research, applications and opportunities. *Materials & Design*, 2014. **56**: 1078–1113.
34. Weertman, J, Steady-state creep through dislocation climb. *Journal of Applied Physics*, 1957. **28**(3): 362–364.
35. Ashby, MF, On interface-reaction control of Nabarro–Herring creep and sintering. *Scripta Metallurgica*, 1969. **3**(11): 837–842.
36. Nix, WD, The effects of grain shape on Nabarro–Herring and Coble creep processes. *Metals Forum*, 1981. **4**(1–2): 38–43.
37. Owen, DM and TG Langdon, Low stress creep behavior: an examination of Nabarro–Herring and Harper–Dorn creep. *Materials Science and Engineering A: Structural Materials Properties Microstructure and Processing*, 1996. **216**(1–2): 20–29.
38. Yue, QZ, L Liu, WC Yang, TW Huang, J Zhang and HZ Fu, Stress dependence of dislocation networks in elevated temperature creep of Ni-based single crystal superalloy. *Materials Science and Engineering A: Structural Materials Properties Microstructure and Processing*, 2019. **742**: 132–137.
39. Chandler, HD, Steady state power law creep resulting from dislocation substructure saturation being approached at higher stress. *Materials Science and Engineering A: Structural Materials Properties Microstructure and Processing*, 2020. **771** 138622, https://doi.org/10.1016/j.msea.2019.138622.

5 Tensile Testing

The uniaxial tensile test is the most commonly used mechanical testing procedure, and indeed it is in very widespread use. However, while it is simple in principle, there are several practical challenges, as well as a number of points to be noted when examining outcomes. For example, there is the issue of converting between nominal ("engineering") and true values of the stress and strain. While many stress–strain curves are presented, and often interpreted, only as nominal data, it is the true relationship that accurately reflects the mechanical response of the sample. Furthermore, conversion between nominal and true values is straightforward only while the stress and strain fields within the gauge length of the sample are uniform. This uniformity is lost as soon as the sample starts to deform in an inhomogeneous way within the gauge length, which most commonly takes the form of "necking." After the onset of necking, which may be quite difficult to detect and could occur at an early stage, useful interpretation of the stress–strain curve becomes difficult. However, FEM modeling does allow various insights into the behavior in this regime, with potential for revealing information (about the fracture event) that is otherwise inaccessible. There are also several important points relating to the way that the strain is measured during a test.

5.1 Specimen Shape and Gripping

5.1.1 Testing Standards

A central issue concerns the specimen size and shape. The behavior is normally monitored in a central section (the "***gauge length***"), in which a uniform stress is created. The ***grips*** lie outside of this section, where the sample has a larger sectional area, so that stresses are lower. If this is not done, then **stress concentration effects** near the grips are likely to result in premature deformation and failure in that area. Several different geometries of sample and grips are possible, with a number of associated standards and protocols. Among the more widely used standards are those of the **American Society of Testing and Materials** (ASTM) and those issued by the **British Standards Institute** (BSI). The latter cover a huge range, but among those of most relevance is BS EN ISO 6892-1:2009 (*Metallic Materials. Tensile Testing. Method of Test at Ambient Temperature*). Among the most common ASTM protocols for tensile testing is ASTM D638. There are, of course, several other types of mechanical test, aimed at extracting different properties (such as fracture toughness), and at testing under different conditions (high strain rate, high temperature etc.).

Even for "conventional" tensile testing at ambient temperature, under quasi-static conditions (relatively low strain rates), there are many aspects of the test that require attention if the results are to be meaningful and universally comparable. The standards are aimed at ensuring this, recognizing that there may sometimes be constraints on the sample dimensions – for example if **thin sheet** is being tested, or if the material is difficult to machine. Full details concerning specifications etc. are available in the standards themselves and also in certain handbooks [1, 2]. The objective in this chapter is to give an overview of the underlying scientific issues and how they relate to sample design, measurement systems (particularly for strain) and interpretation of experimental data.

5.1.2 Geometrical Issues and Stress Fields during Loading

Tensile testing, conducted in a reasonably well-defined way, has been in use for over a century and its application to single crystals also dates back almost as far [3, 4]. More detailed formal study of the theory behind the tensile test also dates back over 50 years [5]. Geometrical issues are clearly important and Fig. 5.1 shows the main options regarding sample shape and gripping systems. As indicated, the **shoulders** of the sample can be manufactured in various ways so as to allow effective gripping. There are pros and cons in each case. For example, shoulders designed for **serrated grips** are easy and cheap to manufacture, but sample alignment can be problematic. On the other hand, a **pinned grip** assures good alignment. **Threaded shoulders** and grips also facilitate good alignment, but the length of the engaged threaded section must be sufficient to take the (maximum) load without any danger of them being stripped. Figure 5.2 shows a photograph of a typical sample and (**split-collar**) gripping system.

There are also issues related to ease of machining and material availability. Most metals can be machined fairly readily, but there may be limitations with both very hard and very soft materials. This is particularly relevant to threaded samples. While there may be options such as **electro-discharge machining** (EDM) [6], and possibly **laser cutting** [7] or **water jet cutting** [8], the production of complex shapes can still present difficulties for some materials. Furthermore, in some cases the material may only be available in limited quantities or in specific shapes. For example, if the interest is in the properties of **thin sheet**, then sample shapes with cylindrical sections are likely to be excluded. (In principle, a circular section would still be possible, by using a very small diameter, but this will result in a sample of small absolute size, which can introduce various difficulties – see §5.1.3.)

The sample shape and gripping geometry also carry implications in terms of the stress field during the test, as well as for sample production. A key objective is to ensure that, in the part of the sample being monitored (the gauge length), the stress field is well defined (uniaxial) and uniform. As described below (§5.4), this condition

5.1 Specimen Shape and Gripping

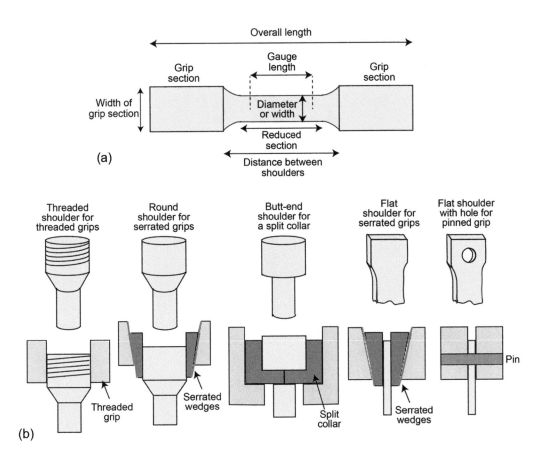

Fig. 5.1 Schematic representations of: (a) a generic sample shape, showing the key dimensions, and (b) the main options regarding types of sample and grip geometry.

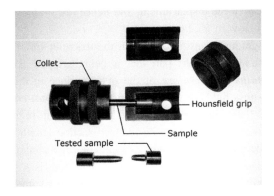

Fig. 5.2 Photograph of a Hounsfield (split collar) gripping system and sample.

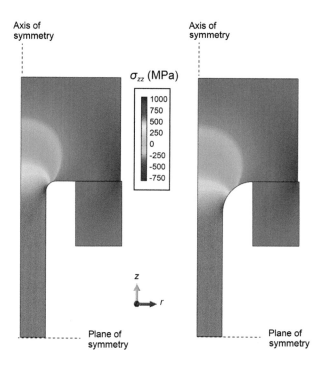

Fig. 5.3 Predicted (elastic) stress fields for loading via a support collar (Hounsfield-type grip), for a root radius of 20% (left) and 60% (right) of the sample diameter.

is likely to be completely lost at some point during the test, but it should at least apply at the start. Ideally, both the load and the displacement associated with the gauge length are accurately monitored throughout. There is likely to be straining (elastic and possibly plastic) outside of the gauge length, but this should not significantly affect the measurements being recorded for the response of the gauge length. This is not usually a problem in terms of the load, which should have the same value throughout all sections of the sample and/or gripping and loading system, but there are several issues associated with measurement of the displacement, and hence the strain, in the gauge length. These are covered below in §5.2.

It's common to have a gradual transition between the diameter (or width) in the "**reduced section length**" and that in the part that is being gripped. This is often defined in terms of the **radius of curvature** of the outer surface in this transition region. Figure 5.3 shows stress fields in (loaded) samples with a ***butt-end shoulder*** (Fig. 5.1(b)), for a sharper (left) and more gradual (right) transition region. In both cases, the stress is fairly uniform in most of the reduced section length, and would certainly be uniform in the gauge length (which would normally start below the bottom of the support collar). However, it can be seen that there is a slight stress concentration at the bottom of the transition region, particularly with the sharper transition case on the left. While this wouldn't affect the plasticity characterization, it could cause ***premature fracture*** at that point.

5.1.3 Issues of Sample Size

There is sometimes a motivation for testing of samples over a range of absolute sizes. This is clearly a different issue from that of shape. The depictions shown in Fig. 5.1 have no scale bar, although standards such as BS EN ISO 6892-1:2009 and ASTM D638 do specify recommended dimensions, which are typically of the order of a few tens of mm for the gauge length and about 10 mm for its diameter. Occasionally there is interest in testing relatively **large samples**. This is often feasible, although of course there could be gripping problems and the sample must fit into the space available in the testing set-up. There is also the issue of the **load capacity** of the testing machine. Most standard machines have a capacity in the range 20–200 kN. For a 10 mm diameter cylindrical sample, with a UTS (peak nominal stress) of, say, 2 GPa (a strong material), the load requirement is about 150 kN. Therefore, using diameters (and lengths) much larger than standard values might well lead to problems with load capacity.

A much more likely scenario is a requirement for testing of **small samples**. This may arise from the inherent nature of the material – for example, it could be in the form of **thin sheet** – or it might be of interest to cut small samples from larger components in order to map spatial variations in properties. (A more common way to do this is to use **hardness testing**, or possibly **indentation plastometry** – see Chapters 7 and 8, but there may be a motivation to use tensile testing for this purpose.) There may also be interest in studying **anisotropy** of components, so that, for example, a tensile sample might be cut with its loading axis transverse to the axis of an extruded bar, constraining the length of the sample. A further possible motivation is the usage of low load capability set-ups, such as those that can be introduced into the chambers of electron microscopes for *in situ* testing. Their usage is often limited to samples of **small cross-section**.

Some of the issues involved are simply practical ones, such as the challenge of accurate strain measurement (§5.2.1) with a **short gauge length**. Others relate to the microstructure of the material and the idea of testing a *"representative volume."* This is an important concept, which also relates to hardness and indentation plastometry. The plasticity characteristics of a material are the outcome of a complex interplay between various microstructural features, which is described in some detail in Chapter 4. The outcome is sensitive to *crystal structure, grain size, crystallographic texture, alloy composition, phase constitution, grain boundary structure, prior dislocation density, impurity levels* etc. For example, the two DIC images shown in Fig. 4.20(b) give a pictorial indication of the importance of interrogating a relatively large (representative) assembly of grains, if the bulk response of the material is to be accurately captured.

It is clear that the presence of a *free surface* close to most of the material being tested is likely to affect its response. The constraint and interactions that normally act between neighboring grains (see §4.3.1) are largely relaxed in that case. These issues, and the dangers associated with testing of samples having small sectional areas, have been investigated, and experimental discrepancies highlighted, although they are not

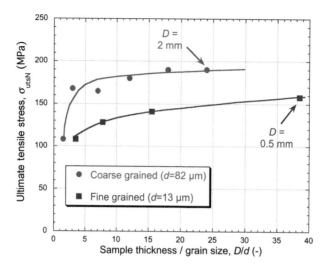

Fig. 5.4 Experimental data [11] from tensile testing of relatively small Cu samples with a range of thicknesses.

universally recognized. One problem is that, if a comparison is made between test outcomes from, say, sheet rolled down to different thicknesses, then there might well be genuine differences in the material properties (due to different degrees of cold work etc.). Only by comparing test outcomes from samples of different size, but identical microstructures, can discrepancies solely due to sample size be identified.

There has been work along these lines [9–11]. For example, Fig. 5.4 shows data [11] from tensile testing of pure Cu samples (with a gauge length of 4 mm and a width of 2 mm). Two types of material were produced, one by *electron beam melting* and annealing (giving a relatively coarse grain size of about 80 µm) and the other by *friction stir processing* (giving a finer grain size of just over 10 µm). Sample thicknesses were varied by grinding and polishing, giving a maximum thickness of 2 mm for the coarse-grained material and 0.5 mm for the fine-grained one. The data plotted in Fig. 5.4 relate to UTS values, although other property values were also reported. It can be seen that these values are sensitive to sample dimensions, being lower for samples with reduced thickness (as a ratio to the grain size). This is due to the reduced constraint that is imposed on the way that the deformation takes place in the very thin samples.

Furthermore, since the minimum ratio required to give "correct" values (independent of this ratio) is higher for the finer-grained material, there is apparently also an absolute size effect, as well as one expressed in terms of the ratio. This is certainly plausible, since constraint reduction effects will operate near to a free surface even if there are many grains across the section. It can thus be concluded that unreliable results are likely to be obtained if there are only a few grains across the section of the sample (even if it is only in one direction) and also if a significant proportion of the section is close enough to a free surface for constraint effects to be relaxed. Unfortunately, this casts doubt on quite a lot of published data.

5.2 Measurement of Load and Displacement

5.2.1 Creation and Measurement of Load

There are two main types of loading frame. One is based on a hydraulic system, with the force being created by the transmission of pressurized oil to pistons. Systems with servo-hydraulic valves are normally used, so that the pressure created is continuously variable. The other type is mechanical, usually involving rotating worm drives (lead screws). In most cases, a cross-head moves between parallel pillars, although some (lower load) systems are based on a simple cantilever arrangement (as are most hardness machines – see Chapter 7). Hydraulic systems are in general better suited to very high loads and to cyclic loading. However, the majority of standard tensile machines are mechanical.

All testing systems have some sort of *"loading train,"* of which the sample forms a part. This "train" can be relatively complex – for example, it might involve a rotating worm drive (lead screw) somewhere, with the force transmitted to a ***cross-head*** and thence via a gripping system to the sample and then to a base-plate of some sort. It does, of course, need to be arranged that, apart from the sample, all of the components loaded in this way experience only elastic deformation. The same force (load) is being transmitted along the complete length of the loading train. Measurement of this load is thus fairly straightforward. For example, a ***load cell*** can be located anywhere in the train, possibly just above the gripping system. Load cells, of which there are several different designs, are in general reliable and accurate, and easy to incorporate into loading systems. Handbooks are available [12] that provide a wealth of technical detail. In some simple systems, such as hardness testers or creep rigs, a ***fixed load*** may be generated by a ***dead weight***.

5.2.2 Displacement Measurement Devices

Measurement of the displacement (in the gauge length), and hence of the strain, presents much more of a challenge than that of monitoring the load. Sometimes, a measuring device is built into the loading frame – for example, it could measure the amount of ***rotation of a lead screw***. In such cases, however, measured displacements include a contribution (elastic) from various elements of the loading train, and this could be quite significant. It may therefore be important to apply a ***compliance calibration***. This involves subtracting from the measured displacement the contribution due to the compliance (inverse of stiffness) of the loading train. This can be measured using a sample of known stiffness (ensuring that it remains elastic). It may even be possible to do this calibration "internally" to the test, using a known Young's modulus of the material.

While such "remote" measurement of displacement, with a compliance calibration, can be acceptable in some cases, it rarely provides very high accuracy. There may be a contribution to the displacement from parts of the sample outside of the gauge length and this is sometimes difficult to subtract. It is often preferable for the displacement to be measured directly on the gauge length, eliminating concerns about the system compliance. Devices of this type include ***clip gauges*** (knife edges pushing lightly into the sample) and ***strain gauges*** (stuck on the sample with adhesive). The latter have good accuracy ($\pm 0.1\%$ of the reading), but are limited in range (~1–2% strain). They are useful for measurement of the sample stiffness (Young's modulus), but not for plastic deformation. A clip gauge is therefore often a good solution. They are widely used in fracture toughness testing, for example to measure crack tip opening displacements, as well as in tensile testing. It's also possible in some cases to use ***linear variable displacement transducers*** (***LVDTs***), although they can't normally be connected directly to the gauge length. Technical details about devices such as LVDTs and clip gauges are available in handbooks [13] on transducers. Typical resolutions of these devices are usually around 1 μm. Of course, the accuracy of the resultant strain value depends on the gauge length, but in general it is likely to be good (better than 0.1%).

There are also a number of non-contacting systems, with reviews [10, 14–16] of these available. A versatile technique, useful for mapping strains over a surface, is ***digital image correlation*** (***DIC***), in which the motion of features ("speckles") in optical images is followed automatically during deformation, with displacement resolutions typically of the order of a few μm. ***Eddy current*** gauges and ***scanning laser extensometers*** can also be used. These also have resolutions of the order of 1 μm. More specialized (and accurate) devices include ***parallel plate capacitors*** and ***interferometric optical*** set-ups, although they often have more limited measurement ranges.

5.3 Tensile Stress–Strain Curves

5.3.1 Nominal and True Plots

The standard outcome of a tensile test is a plot of the nominal stress against the nominal strain. As described in §3.1.3, it is straightforward to convert between true and nominal (engineering) values of stress and strain, using simple analytical equations, although it is important to understand that these conversions are only valid if the stress and strain fields within the sample (gauge length) are uniform (homogeneous) – which can only be true prior to the onset of necking. In practice, it is quite common to present only the nominal plot, and several common procedures for extraction of key parameters are based only on inspection of such curves. However, if the objective is to obtain fundamental information about the plasticity (and failure) characteristics of the material, then it is a plot of true stress against true strain that provides this. Furthermore, the important phenomena of neck formation and subsequent growth can only be understood once the relationship between the two types of stress and strain has been fully grasped.

5.3.2 The Onset of Necking and Considère's Construction

With a brittle material, tensile testing may give an approximately linear stress–strain plot, followed by fracture (at a stress that may be affected by the presence and size of flaws). However, most metals do not behave in this way and are likely to experience considerable plastic deformation before they fail. Initially, this is likely to be uniform throughout the gauge length. Eventually, of course, the sample will fail (fracture). However, in most cases, failure will be preceded by at least some **necking**. The formation of a neck is a type of instability, the formation of which is closely tied in with work hardening. Once a neck starts to form, the (true) stress there will be higher than elsewhere, probably leading to more straining there, further reducing the local sectional area and accelerating the effect.

In the complete absence of work hardening, the sample will be very susceptible to this effect and will be prone to necking from an early stage. (This will be even more likely in a real component under load, where the stress field is likely to be inhomogeneous from the start.) Work hardening, however, acts to suppress necking, since any local region experiencing higher strain will move up the stress–strain curve and require a higher local stress in order for straining to continue there. Generally, this is sufficient to ensure uniform straining and suppress early necking. However, since the work hardening rate often falls off with increasing strain (see §3.2.1 and §4.2.2), this balance is likely to shift and may eventually render the sample vulnerable to necking. Furthermore, some materials (with low work hardening rates) may indeed be susceptible to necking from the very start. It may be noted at this point that work hardening rates are often reduced if the temperature is raised – see §4.2.3. This is a key reason for carrying out various *forming processes (forging, rolling* etc.) at high temperature, allowing large strains to be created without the stresses getting too high. However, if tensile stresses develop during such a process, then necking is likely. Mainly for this reason, most such forming procedures involve only compressive (normal) stresses.

Instabilities of this general type are actually quite common: such a situation was originally analyzed by Armand Considère (1885) in the context of the stability of structures such as bridges. Instability (onset of necking) is expected to occur when an increase in the (local) strain produces no net increase in the load, F. This will happen when

$$\Delta F = 0 \tag{5.1}$$

which leads to

$$F = A\sigma_T, \therefore dF = A\, d\sigma_T + \sigma_T dA = 0$$

$$\therefore \frac{d\sigma_T}{\sigma_T} = -\frac{dA}{A} = \frac{dA}{A} = d\varepsilon_T \tag{5.2}$$

$$\therefore \sigma_T = \frac{d\sigma_T}{d\varepsilon_T}$$

with the T subscript being used to emphasize that these stresses and strains must be true values. Necking is thus predicted to start when the slope of the true stress–true

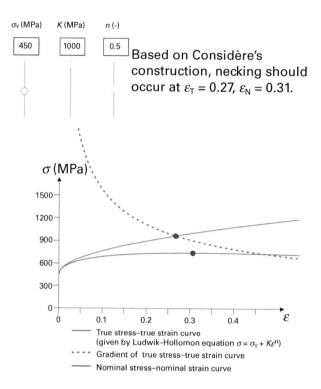

Fig. 5.5 The Considère construction for prediction of the onset of necking, expressed in the form of Eqn. (5.2) and applied to a material exhibiting a Ludwik–Hollomon true stress–true strain curve, with the parameter values shown.

strain curve falls to a value equal to the true stress at that point. Figure 5.5 shows the construction for a (true) stress–strain curve represented by a simple analytical expression (Eqn. (3.13)). This plot was obtained from a DoITPoMS interactive educational software package – see www.doitpoms.ac.uk/tlplib/mechanical_testing_metals/necking.php.

The condition can also be expressed in terms of the nominal strain:

$$\frac{d\sigma_T}{d\varepsilon_T} = \frac{d\sigma_T}{d\varepsilon_N}\frac{d\varepsilon_N}{d\varepsilon_T} = \frac{d\sigma_T}{d\varepsilon_N}\left(\frac{dL/L_0}{dL/L}\right) = \frac{d\sigma_T}{d\varepsilon_N}\left(\frac{L}{L_0}\right) = \frac{d\sigma_T}{d\varepsilon_N}(1+\varepsilon_N)$$

Therefore, at the instability point,

$$\sigma_T = \frac{d\sigma_T}{d\varepsilon_N}(1+\varepsilon_N) \tag{5.3}$$

It can therefore also be formulated in terms of a plot of true stress against nominal strain. On such a plot, necking will start where a line from the point $\varepsilon_N = -1$ forms a tangent to the curve. This is shown in Fig. 5.6, which was obtained using the same Ludwik–Hollomon representation of the true stress–true strain relationship as that of Fig. 5.5.

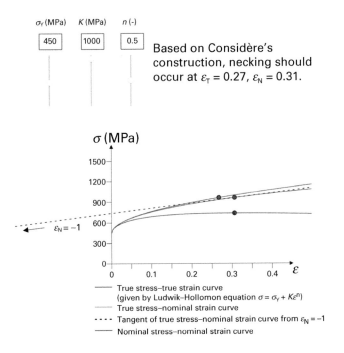

Fig. 5.6 The Considère construction, expressed in the form of Eqn. (5.3) and applied to a material exhibiting a Ludwik–Hollomon true stress–true strain curve, with the parameter values shown.

It's important to note that the condition also corresponds to a peak (plateau) in the nominal stress–nominal strain plot. This can be seen on obtaining the gradient of such a plot by differentiating the expression for σ_N (Eqn. (3.10)) with respect to ε_N, as follows:

$$\sigma_N = \frac{\sigma_T}{(1+\varepsilon_N)}$$

$$\therefore \frac{d\sigma_N}{d\varepsilon_N} = \frac{d\sigma_T}{d\varepsilon_N}\frac{1}{(1+\varepsilon_N)} - \frac{\sigma_T}{(1+\varepsilon_N)^2}$$

Substituting for the true stress–nominal strain gradient (at the onset of necking) from Eqn. (5.3):

$$\frac{d\sigma_N}{d\varepsilon_N} = \frac{\sigma_T}{(1+\varepsilon_N)}\frac{1}{(1+\varepsilon_N)} - \frac{\sigma_T}{(1+\varepsilon_N)^2} = 0 \tag{5.4}$$

This condition can also be seen in Figs. 5.5 and 5.6. Since many stress–strain curves are presented as nominal plots, and this is a simple condition that can be identified by visual inspection, it is in many ways the easiest criterion to use in order to establish the onset of necking. It also corresponds to the "strength" (***ultimate tensile stress*** – see §5.3.3 below).

However, it should be noted that the "plateau" is often a fairly flat one, as in the plots of Figs. 5.5 and 5.6. If there is interest in the strain at which necking starts, then this may in practice be a rather inaccurate approach and it may be preferable to use the construction shown in Fig. 5.5 – i.e. to use Eqn. (5.2), after converting the nominal stress–strain curve to a true one (using Eqns. (3.10) and (3.11)). The tangent construction of Eqn. (5.3), shown in Fig. 5.6, while popular for analysis of the *tensile drawing of polymers* [17–19] (since it allows study of the regime of *stable necking*), is rarely used in interpreting the stress–strain curves of metals.

5.3.3 Neck Development, UTS, Ductility and Reduction in Area

The Considère construction has been successfully used for decades to explain the main features of necking (unstable for metals, but stable for some polymers). However, it is not a full description of what happens. For example, it takes no account of the geometry of the sample: even for a radially symmetric shape (cylindrical section), the precise way that the neck forms turns out to be (slightly) dependent on the *aspect ratio* – i.e. the ratio of the uniform (reduced) section length to its diameter. Considère takes no account of any geometrical factors. Furthermore, the treatment offers nothing concerning the *post-necking behavior*. The complete tensile test can, however, be simulated via FEM modeling.

It may be noted at this point that it is common during tensile testing to extract the "*strength*," in the form of an "*ultimate tensile stress*" (***UTS***). This is usually taken to be the peak on the nominal stress v. nominal strain plot, which corresponds to the onset of necking, as outlined above. It should be understood that this value is not actually the true stress acting at failure. This is difficult to obtain in a simple way, since, once necking has started, the (changing) sectional area is unknown – although the behavior can often be captured quite accurately via FEM modeling – see below. Also, the "*ductility*" (or "*failure strain*," or "*elongation at failure*"), often taken to be the nominal strain when fracture occurs, which is usually well beyond the strain at the onset of necking, does not correspond to the true strain in the neck when fracture occurs. In fact, the values quoted for ductility have little or no real significance, despite their widespread usage.

This point is illustrated by the plots [20] shown in Fig. 5.7, which relate to a single material that was tensile tested with a range of values for the gauge length, which in this case was also the length of the reduced section part of the sample, and the diameter of the gauge section (which was circular). It may be noted here that, while sample dimensions are affecting the outcome, the effect is quite different from those described in §5.1.3, since none of these samples were "small," or had relatively few grains across the section. It can be seen that, while the behavior was similar for all samples up to the point of necking (peak in the plot), which was at about 6–8% strain for this material, the elongation to failure values cover a huge range, being larger for the samples with shorter gauge length and, with a given gauge length, for those with larger diameter (i.e. for those with a small aspect ratio).

The cause of this is simple. After the peak, with necking taking place, virtually all of the recorded elongation is due to straining in the neck. For shorter samples, this region constitutes a greater proportion of the gauge length, making the increase in

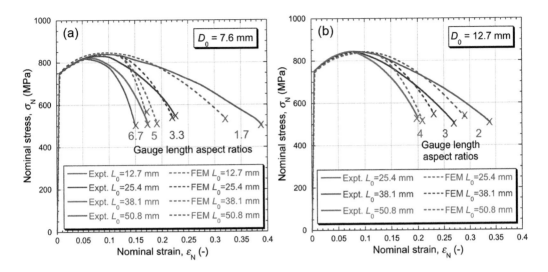

Fig. 5.7 Experimental plots [20], and corresponding FEM predictions, of nominal stress against nominal strain, from tensile testing of HY-100 steel, with the samples having a range of values for the gauge length (L_0) and diameter of the gauge section (D_0). The aspect ratios (L_0/D_0) are also indicated. The FEM modeling is based on the Voce law (Eqn. (3.14)), with $\sigma_Y = 740$ MPa, $\sigma_s = 1035$ MPa and $\varepsilon_0 = 0.1$.

(nominal) "strain" larger. Similarly, with a larger diameter, the contribution from necking is increased (for a given gauge length). This effect can be clearly captured in an FEM model, as shown in Fig. 5.7. Provided that the (true) stress–strain relationship is valid up to the high strains involved, and a suitable fracture criterion can be identified – a true (von Mises) plastic strain of 100% was used in the plots shown, then the complete (nominal) stress–strain curve, including the post-necking region, can be reliably predicted.

Of course, such modeling is not routinely undertaken and these features highlight the fact that, certainly in an absolute sense, "ductility" values are virtually meaningless. The actual (true) strain in the neck at the point of fracture bears no direct relation to the raw number obtained from the nominal stress–strain curve – the true strain in the neck is often considerably higher. Also, the true stress at the point of fracture is usually higher than the apparent value according to the plot. The load often drops while the neck develops, but the sectional area in the neck is also dropping (more sharply), so the true stress there is rising. Again, there is no simple way of estimating this value, since it depends on the geometry of the neck.

It should also be noted that there is another parameter that is commonly extracted from a tensile test, which is the so-called "***reduction in area***" (***RA***). This is defined as the decrease in sectional area at the neck (usually obtained by measurement of the diameter at one or both of the fractured ends), divided by the original sectional area. It is sometimes stated that this is a more reliable indicator of the "ductility" than the elongation at failure (partly in recognition of the fact that the latter is dependent on the aspect ratio of the gauge length, although this dependence is far from being universally acknowledged). There is something in this argument, but the RA is still some way from being a genuinely meaningful parameter. One objection is that it is not easy to measure accurately, particularly with samples that are not circular in section. Rather more fundamentally, it is affected by both the uniform plastic deformation that took place before necking and by the development of the neck. Furthermore, it is sensitive to exactly what happens in the latter stages of necking, when the true strain is often becoming very high and the behavior is of limited significance in terms of a meaningful definition of strength (or toughness). The above points are in fact relatively simple and have been extensively recognized and studied [21–26], frequently via FEM modeling.

It is sometimes stated that the initiation of necking during tensile testing arises from (small) variations in sectional area along the gauge length of the sample. However, in practice, for a particular metal, its onset does not depend on whether great care has been taken to avoid any such fluctuations. Furthermore, the introduction of such defects in an FEM model does not, in general, significantly affect the predicted onset. The (modeling) condition that does lead to necking is the assumption that, near the end of the gauge length, the sample is constrained from contracting laterally [22, 27, 28]. In practice, due to the increasing sectional area in that region, and because the material beyond the reduced section length will undergo little or no deformation, that condition is often a fairly realistic one.

In this way, for any true stress–true strain relationship, including an experimental one that cannot be expressed as an equation, FEM simulation can be used to predict the onset of necking. Modeling outcomes are shown here for copper in two states, with the Voce law being used to represent the (true) stress–strain relationship in both cases. The parameter values used are typical of annealed and as-received (work-hardened) copper [29], with the former showing high work hardening capacity and the latter much less. The corresponding stress–strain relationships (true and nominal) are plotted in Fig. 5.8, with the nominal curves obtained from the Voce equation by simple use of Eqns. (3.10) and (3.11). The Considère construction was used to identify the onset of necking points, although, as can be seen in Fig. 5.7, both experiment and FEM simulation reveal that it actually has a dependence on the aspect ratio of the sample.

These constitutive relationships were used in the FEM simulations of Figs. 5.9 and 5.10. Firstly, Fig. 5.9 gives a pictorial indication of the state of these two samples soon after the onset of necking. This figure shows both photos of the samples and FEM-predicted sample shapes and fields of (axial) stress and strain within them. The AR-Cu, which exhibits little work hardening, necks at a nominal strain of about 15–20%. The stress and strain fields shown are starting to exhibit marked inhomogeneity, with levels of both rising in the neck region. The Ann-Cu exhibits more work hardening, resulting in a delay of necking up to about 30–35%.

Of course, there is interest in the level of agreement between experimental and modeled stress–strain curves, with the comparison commonly made between nominal plots. Such a comparison is shown in Fig. 5.10, for these two materials. As for

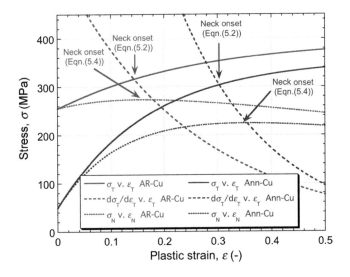

Fig. 5.8 Stress–strain plots, for as-received (AR) and annealed (Ann) copper, conforming to the Voce equation (Eqn. (3.14)), with parameter value sets of $\sigma_s = 395$ MPa, $\sigma_Y = 255$ MPa and $\varepsilon_0 = 0.25$ (AR-Cu) and $\sigma_s = 355$ MPa, $\sigma_Y = 49$ MPa and $\varepsilon_0 = 0.17$ (Ann-Cu).

Fig. 5.7, it is clear that the main features are well captured for both of them. A note should be made here concerning the failure criterion. A critical (von Mises) strain level was again used, with a value of 70% for the AR-Cu and 50% for the Ann-Cu. It may be helpful to note here the relationship between the axial strains in Fig. 5.9 and the von Mises strain used as the failure criterion. The von Mises strain, in terms of the principal strains, is given by the analogous expression to that for stress (Eqn. (3.3)):

$$\varepsilon_{vM} = \sqrt{\frac{(\varepsilon_1 - \varepsilon_2)^2 + (\varepsilon_2 - \varepsilon_3)^2 + (\varepsilon_3 - \varepsilon_1)^2}{2}} \tag{5.5}$$

In this case, the axial strain approximates to a principal strain (at least until the neck becomes heavily distorted) and the two transverse strains will (both) be whatever is needed to conserve volume. For example, an axial strain of 50% will mean that the other two are both -25% (since $\varepsilon_1 + \varepsilon_2 + \varepsilon_3 = 0$, see Eqn. (2.21)) and ε_{vM} is 75%. Critical von Mises strains of 70% and 50% thus correspond to axial strains of about 47% and 34%. This is broadly consistent with the strain fields in Fig. 5.9, noting that failure occurred well past the necking point in the AR-Cu, but close to it for the Ann-Cu.

It can also be seen that the FEM-predicted points for the onset of necking in these two cases (~15–20% and ~35–40% for AR-Cu and Ann-Cu, respectively) are close to (slightly above) those from Considère (Fig. 5.8). As mentioned above, Considère takes no account of the constraint conditions, or the aspect ratio of the sample, in the way that the FEM model does. The FEM prediction of the necking strain is expected to be slightly more reliable, although, for most purposes, Considère is acceptable. The

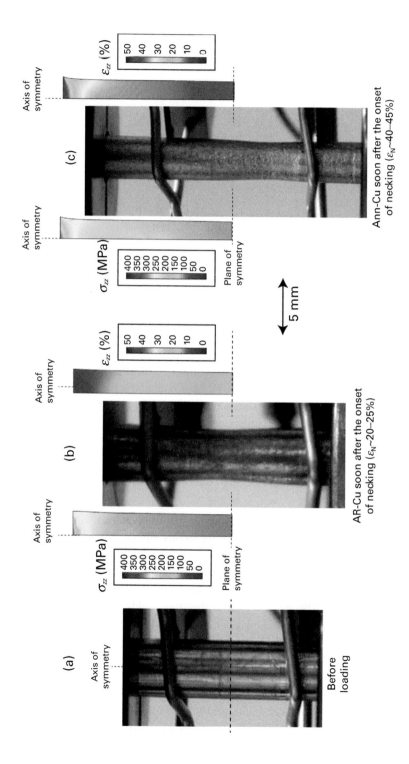

Fig. 5.9 Progression of the tensile test for the two coppers of Fig. 5.8, showing photos and FEM-predicted fields of stress and strain (a) before loading (no stress or strain), (b) the AR-Cu, soon after necking onset, and (c) the Ann-Cu, soon after necking onset.

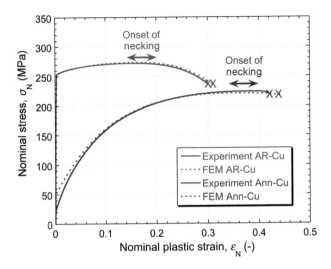

Fig. 5.10 Comparison between the (nominal) stress–strain plots for two copper materials (as-received and annealed), obtained experimentally and via FEM modeling (using the Voce plasticity parameter values in the caption of Fig. 5.8, a reduced section length of 30 mm, a gauge length of 12.5 mm, a diameter of 5 mm and critical strains to failure of 70% and 50%, respectively).

neck will always be predicted to form at the mid-point of the sample, but the differences introduced if it forms elsewhere are very small (provided it is within the gauge length).

There is thus scope for using FEM (with an appropriate true stress–strain relationship) to predict the complete tensile stress–strain curve, including the necking and rupture, but a caveat should be added. Such predictions are based on assuming that the (true) stress–strain relationship holds up to the (high) strains that are likely to be generated in the neck. Since this relationship will have been inferred only on the basis of the response up to the onset of necking (perhaps a few tens of % at most), and the strains created in the neck may reach values of the order of 100%, this may not be reliable. Moreover, it's not just an issue of extrapolating a curve well beyond the measurable regime, since the behavior may change at these high strain levels. For example, cavitation could occur in the neck shortly before final fracture, which might affect the stress–strain relationship.

It may be noted here that the indentation plastometry technique (Chapter 8) offers potential advantages over tensile testing in this respect, since it's often possible to create significantly higher plastic strains (in a controlled way) during indentation, so that the inferred stress–strain relationship can be representative of the behavior over a greater range of plastic strain than that created (in a well-defined way) during tensile testing. (This operation of inferring the true stress–strain relationship involves inverse, or iterative, FEM simulation, which is fully described in Chapter 8.)

In summary, while the concept of the UTS is of at least some significance and value, particularly if considered in combination with the (true) stress–strain relationship during plastic deformation, the numbers obtained for the elongation at failure and the RA are more or less meaningless. There is a strong argument for abandoning them entirely and concentrating on obtaining parameters that provide useful guidelines for assessment of the "strength" of a metal. On the other hand, provided the stress–strain relationship can be well-captured in a constitutive law (that holds up to relatively high strains), FEM simulation can be used to predict the complete (nominal) stress–strain curve. By comparing simulated and experimental curves in terms of the point at which fracture occurs, it may be possible to estimate the critical (von Mises) strain for fracture. This is a parameter that is widely used in FEM simulation of various practical situations, so being able to evaluate it for a particular material in this way is an attractive concept.

5.3.4 Load Drops and Formation of Lüders Bands

There is a common perception that certain materials, notably **mild steel**, can exhibit a stress–strain curve of the type shown schematically in Fig. 4.25(a), with a load drop appearing at the end of the elastic regime, followed by a burst of plastic straining at constant load. The traditional explanation for this behavior is depicted in Fig. 4.24, which shows how dislocations can become **pinned** by (interstitial) C atoms. When the applied stress becomes sufficiently high, these dislocations escape from their "atmospheres," with release of energy, allowing them to move relatively large distances (creating a burst of plasticity with no further increase in the load). Conventional work hardening then follows, although (in certain regimes of temperature and strain rate) "serrations" may subsequently appear in the stress–strain curve (corresponding to repeated atmosphere reformation and dislocation escape), termed the ***Portevin–Le Chatelier effect*** – see Fig. 4.25(b).

This burst of plasticity without any rise in load renders the material susceptible to localized plasticity, which is normally opposed by the work hardening effect – see §5.3.2 and §5.3.3. Indeed, materials that exhibit a stress–strain curve of this type are likely to exhibit such local plastic straining, although it does not become unstable in the way that necking does, since the extent of this burst of plasticity is quite limited – usually of the order of 2%. The form that this plasticity takes is often that of a narrow "band" running across the sample at an angle to the loading axis. This is termed a "***Lüders band***," or possibly a "***Chernov–Lüders band***" (commonly by Russian authors). These bands sometimes propagate along the length of the sample, and may or may not be apparent in the final deformed sample. This is more likely if the deformation has only been taken to relatively low levels of plastic strain. They can be seen in steel sheets that have been formed in various ways. They are sometimes referred to as "***stretcher strains***."

The literature concerning Lüders bands is extensive and potentially rather confusing. The original attribution is a little unclear, but scientific attention started to focus on the effect after the work of Cottrell and Bilby [30], who introduced the concept of ***dislocation atmospheres*** in the late 1940s. Since then, study of Lüders band formation has been extensive. Some such work has clearly related to slightly different

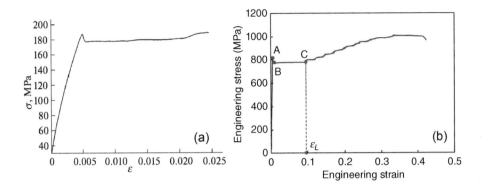

Fig. 5.11 Experimental stress–strain plots for (a) 08PS steel [36] and (b) a medium-Mn TRIP steel [39].

phenomena. For example, the label has been attached [31, 32] to well-established effects in single crystals, where there is no load drop, the initial straining without a rise in stress is just the "easy glide" regime of operation of a single slip system (see §4.2.1 and Fig. 4.11), and the "bands" that are seen in the sample are just steps created where sets of dislocations reach the free surface ("*persistent slip bands*"). There are also various reports [33, 34] of "Lüders band formation" in materials such as Al alloys and brasses, but without any evidence of load drops or any indication of how they might have formed (in systems without interstitials and hence without any expectation of strong dislocation pinning, at least in the way envisaged in the concept of Lüders bands).

Nevertheless, it is clear that, at least under some conditions and with certain materials (mainly steels), a load drop is observed, followed by a short burst of plastic straining, attributable to dislocation pinning (by interstitial atoms) and subsequent escape, and this can lead to narrow bands of localized plasticity in the sample [35–38]. Examples of experimental stress–strain curves of this type are shown in Fig. 5.11. Figure 5.11(a) relates to 08PS steel [36], which is a structural carbon steel (with about 0.1%C, as well as some Si and Mn). It's certainly not highly alloyed, but neither is it high purity Fe, which is sometimes perceived as the only material in which the effect is very clear. (The term "*mild steel*" is not very well defined.)

There are also reports of Lüders band formation in steels that are quite different from this. For example, it has been reported [39–41] in TRIP (*transformation-induced plasticity*) steels, with compositions such as Fe−7Mn−0.14C−0.23Si. These are much harder materials than "mild steel," with the hardening attributable to *retained austenite* (stabilized by the Mn) being transformed to martensite on loading. It seems unlikely that the carbon has much of a role in dislocation pinning. Nevertheless, it can be seen in Fig. 5.11(b) that there is a clear plateau after yielding (up to a high strain of about 10%), although there's little or no yield drop. It's also rather unclear whether well-defined bands are left in the sample. There is certainly evidence of the propagation along the sample during the test of a region of localized straining, picked up both via optical monitoring of the complete strain field and from the associated heating effect [39]. There is thus an argument for classifying this as a Lüders-type phenomenon. In fact, the thermal aspect of the process in this case is worth noting, since it's possible that local heating plays a role in promoting strain localization [39, 42]. There has also been work [43] involving the monitoring of Lüders band propagation from the associated acoustic emissions.

5.4 Variants of the Tensile Test

5.4.1 Testing of Single Crystals

Several points should be noted about the testing of single crystals, compared with (the much more common) case of polycrystalline samples. Some of these are covered in §4.2.1 and §4.3.1. As highlighted there (e.g. Fig. 4.20), the stress–strain response tends to be very different in the two cases, with single crystals exhibiting an ***anisotropic*** response, and often deforming plastically at lower stress levels (although this may not be true for creep deformation). Furthermore, the rate of work hardening commonly increases with increasing strain (as multiple slip starts to operate), which is unusual for polycrystals.

The anisotropic response of single crystals has several consequences. Of course, the response of textured polycrystals can also be anisotropic. However, at least for most testing scenarios, such samples are ***transversely isotropic*** during any particular test. With single crystals, on the other hand, there is often a strong tendency for the sample to shear transversely to the testing axis (as shown in Fig. 4.10), at least during the initial easy glide regime. This leads to the outcome being quite sensitive to the exact nature of the grip constraints, an effect that has been clear for a considerable time [44]. Of course, there is also a dependence on the crystal structure and the initial orientation of the tensile axis. Kim and Greer reported [45] significant ***tensile-compressive asymmetry*** (see §6.3.1) during testing of Mo (bcc) single crystals, but not with Au (fcc). They explain this in terms of dislocation dynamics and sample constraint, although it should be noted that their samples were all very small (<1 μm diameter), which probably introduced size effects of the type described in §5.1.3.

5.4.2 Biaxial Tensile Testing

Most tensile testing is done under simple uniaxial loading. However, ***biaxial loading*** tests are sometimes carried out, usually on samples cut from ***metal sheet***. It's important to understand the potential motivation for such testing, which tends to be experimentally more complex (usually involving ***cruciform samples***). Of course, there is more scope for controlling the stress state in the sample, although the through-thickness stress will still normally be zero (since the top and bottom of the sheet are ***free surfaces***, at which there can be ***no normal stress***). Therefore, according to the ***Tresca yield criterion*** (§3.1.2), if both applied stresses are tensile, then the onset of plasticity will take place when the larger of the two reaches the uniaxial yield stress, σ_Y, with the other applied stress having no effect. This is not, however, true with the **von Mises yield criterion** (§3.1.2), for which all three of the principal stresses are predicted to have an effect, so this kind of testing can be used to check which of the two criteria is more reliable. (Under uniaxial loading, both criteria predict that yielding will occur when the applied stress reaches σ_Y – see Fig. 3.3.)

More commonly, however, this type of testing is used to check on in-plane **anisotropy**. Such anisotropy is in fact quite common in metal sheet (and also in other **formed components**, such as **extruded rod** or **drawn wire**). It is predominantly associated with **crystallographic texture** (non-random distribution of the orientation of individual grains) that arises during deformation processing (or solidification processing). While there are relationships between the nature of the texture and the associated anisotropy in mechanical properties, these are in general quite complex [46–49]. Furthermore, the consequences of (in-plane) anisotropy (in sheet metal) are of practical importance – for example in reducing the incidence of "*earing*" in **deep-drawn sheet** [50–52]. Experimental assessment is therefore important. Work of this type has been oriented towards optimizing test conditions [53, 54] and obtaining yield envelopes that reflect the anisotropy of the sheet [55].

5.4.3 High Strain Rate Tensile Testing

As described in §3.2.2, the plasticity characteristics of a metal can start to change if strain is imposed at a high rate. The mechanistic explanation for this, which relates to limitations on the effective speed of dislocation glide, is briefly outlined in §4.4.1 and §4.4.2. Typically, this effect only starts to become significant at strain rates of $\sim 10^3$ s^{-1} and above. Such rates are commonly generated during various kinds of impact and ballistic event, and there are specialized testing techniques in which they can be created – see §10.4. However, as described in slightly more detail in §10.4.1, it can readily be seen that they are not easily created during conventional tensile testing. For example, with a gauge length of, say, 50 mm, one of the gripped ends will need to be moving (relative to the other) at a speed of about 50 m s^{-1} in order to create such a strain rate. Such speeds are high and of course they need to be attained very quickly, so the acceleration required is likely to be extreme. In fact, even using specialized equipment (see §10.4.1), the strain rate limit with conventional (uniaxial) geometry is usually considered to be a few hundred s^{-1}. Arrangements [56–58] for testing at higher rates than this are mostly variants of the **Split Hopkinson Bar** test (§10.4.2), although the geometry and gripping arrangements are different from that of a conventional tensile test.

A point that should be noted for all high strain rate testing is that the temperature of the sample may rise during the test [59]. That this is expected can readily be demonstrated, since virtually all of the work done during plastic deformation is converted into heat. The temperature rise, for a situation in which none of this heat is lost to the surroundings (as expected when the deformation takes place very quickly), can be obtained by setting the plastic work equal to the thermal energy:

$$\int \sigma \varepsilon_p \, d\varepsilon_p = c \Delta T \tag{5.6}$$

where the left-hand side is the plastic work per unit volume and c is the volume specific heat. If work hardening is ignored for this purpose, then the temperature rise can be written as

$$\Delta T = \frac{\sigma_Y \varepsilon_p}{c} \tag{5.7}$$

For a typical yield stress of 200 MPa and a plastic strain of, say, 50%, given that the value of c is usually of the order of 1 MJ m^{-3} K^{-1}, this indicates a temperature rise of about 100 °C. Clearly such a change could affect the mechanical response, so this effect should be borne in mind. A dependence on temperature is taken into account in some constitutive laws covering the strain rate sensitivity of plasticity, such as that of Johnson–Cook – see §3.2.2.

5.4.4 Tensile Creep Testing

Tensile creep testing is carried out in a similar way to conventional (stress–strain) testing, but there are additional challenges. For example, it is commonly carried out at high temperature, so a furnace (with good thermal stability) is needed, and all of the sample must be held at the selected temperature. Also, the load must be sustained for long periods – perhaps just a few hours, but in some cases periods of many weeks or even many months might be needed. Such conditions bring slightly different challenges from those of conventional testing.

Many creep facilities are based on a fixed load, applied via a **_dead weight_** (with a lever arrangement such that the actual applied force is considerably larger than the weight itself). A typical facility of this type is shown in Fig. 5.12(a). The actual

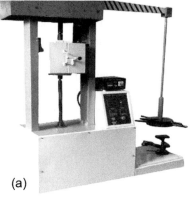

Fig. 5.12 Photos of typical tensile creep testing set-ups, showing (a) fixed weight and (b) variable load machines.

weight used is unlikely to be much more than a few tens of kg. However, with the lever arm arrangement giving a mechanical advantage of at least 10, and perhaps considerably more, such weights can produce an applied force on the sample of up to several tens of kN. This is usually sufficient for most situations (sample dimensions and required stress levels), although the limitations associated with having a *fixed nominal stress* (varying true stress) should be noted. As described in §3.2.4, this effect can make interpretation of the experimental data a little difficult. For example, the so-called "tertiary" regime of creep, in which the strain rate starts to accelerate, may in fact be a consequence of the true stress rising as the sectional area of the sample decreases. This point is not always fully appreciated. There is no such issue with *indentation creep plastometry* (§8.4.4), in which the true stress field is being constantly monitored (modeled) throughout the process.

It is, of course, possible to have a loading frame in which the force can be varied during the test. A typical facility of this type is shown in Fig. 5.12(b). Such machines can readily generate forces of up to hundreds of kN. With the strain (extension) being continuously monitored, software control can be used to change the applied load such that a constant true stress is maintained. However, it is worth noting that such machines tend to be expensive. They're not well suited to very long term tests, both because of potential wear on the loading system and because tying up such expensive facilities for long periods is not economically attractive. If a number of samples are to be tested over periods of weeks or months, which is not unusual for creep testing, then it is more likely that a set of dead weight machines will be used.

References

1. Kuhn, H and D Medlin, *ASM Handbook Vol. 8: Mechanical Testing and Evaluation*. Materials Park, OH: ASM International, 2000.
2. Davis, JR, *Tensile Testing*. Materials Park, OH: ASM International, 2004.
3. Von Goler, F and G Sachs, Tensile tests on crystals of copper and alpha-brass. *Zeitschrift Fur Physik*, 1929. **55**(9–10): 581–620.
4. Osswald, E, Tensile tests on copper, nickel crystals. *Zeitschrift Fur Physik*, 1933. **83**(1–2): 55–78.
5. Hart, EW, Theory of tensile test. *Acta Metallurgica*, 1967. **15**(2): 351–355.
6. Nahak, B and A Gupta, A review on optimization of machining performances and recent developments in electro discharge machining. *Manufacturing Review*, 2019. **6**.
7. Nagimova, A and A Perveen, A review on laser machining of hard to cut materials. *Materials Today: Proceedings*, 2019. **18**: 2440–2447.
8. Kartal, F, A review of the current state of abrasive water-jet turning machining method. *International Journal of Advanced Manufacturing Technology*, 2017. **88**(1–4): 495–505.
9. Simons, G, C Weippert, J Dual and J Villain, Size effects in tensile testing of thin cold rolled and annealed Cu foils. *Materials Science and Engineering A: Structural Materials Properties Microstructure and Processing*, 2006. **416**(1–2): 290–299.
10. Zhao, YH, YZ Guo, Q Wei, TD Topping, AM Dangelewicz, YT Zhu, TG Langdon and EJ Lavernia, Influence of specimen dimensions and strain measurement methods on tensile

stress–strain curves. *Materials Science and Engineering A: Structural Materials Properties Microstructure and Processing*, 2009. **525**(1–2): 68–77.
11. Yang, L and L Lu, The influence of sample thickness on the tensile properties of pure Cu with different grain sizes. *Scripta Materialia*, 2013. **69**(3): 242–245.
12. *Load Cell and Weigh Module Handbook*, 2020. Available from: www.ricelake.com/lcwm.
13. Boyle, HB, *Transducer Handbook*. Oxford: Butterworth-Heinemann, 1992.
14. Bastias, PC, SM Kulkarni, KY Kim and J Gargas, Noncontacting strain measurements during tensile tests. *Experimental Mechanics*, 1996. **36**(1): 78–83.
15. Anwander, M, BG Zagar, B Weiss and H Weiss, Noncontacting strain measurements at high temperatures by the digital laser speckle technique. *Experimental Mechanics*, 2000. **40**(1): 98–105.
16. Pan, B and L Tian, Advanced video extensometer for non-contact, real-time, high-accuracy strain measurement. *Optics Express*, 2016. **24**(17): 19082–19093.
17. McKinley, GH and O Hassager, The Considere condition and rapid stretching of linear and branched polymer melts. *Journal of Rheology*, 1999. **43**(5): 1195–1212.
18. Crist, B and C Metaxas, Neck propagation in polyethylene. *Journal of Polymer Science Part B: Polymer Physics*, 2004. **42**(11): 2081–2091.
19. Petrie, CJS, Considere reconsidered: necking of polymeric liquids. *Chemical Engineering Science*, 2009. **64**(22): 4693–4700.
20. Matic, P, GC Kirby and MI Jolles, The relation of tensile specimen size and geometry effects to unique constitutive parameters for ductile materials. *Proceedings of the Royal Society of London Series A: Mathematical and Physical Sciences*, 1988. **417**(1853): 309–333.
21. Havner, KS, On the onset of necking in the tensile test. *International Journal of Plasticity*, 2004. **20**(4–5): 965–978.
22. Kim, HS, SH Kim and WS Ryu, Finite element analysis of the onset of necking and the post-necking behaviour during uniaxial tensile testing. *Materials Transactions*, 2005. **46**(10): 2159–2163.
23. Joun, M, I Choi, J Eom and M Lee, Finite element analysis of tensile testing with emphasis on necking. *Computational Materials Science*, 2007. **41**(1): 63–69.
24. Choung, JM and SR Cho, Study on true stress correction from tensile tests. *Journal of Mechanical Science and Technology*, 2008. **22**(6): 1039–1051.
25. Osovski, S, D Rittel, JA Rodriguez-Martinez and R Zaera, Dynamic tensile necking: influence of specimen geometry and boundary conditions. *Mechanics of Materials*, 2013. **62**: 1–13.
26. Ho, HC, KF Chung, X Liu, M Xiao and DA Nethercot, Modelling tensile tests on high strength S690 steel materials undergoing large deformations. *Engineering Structures*, 2019. **192**: 305–322.
27. Samuel, EI, BK Choudhary and KBS Rao, Inter-relation between true stress at the onset of necking and true uniform strain in steels – a manifestation of onset to plastic instability. *Materials Science and Engineering A: Structural Materials Properties Microstructure and Processing*, 2008. **480**(1–2): 506–509.
28. Guan, ZP, Quantitative analysis on the onset of necking in rate-dependent tension. *Materials & Design*, 2014. **56**: 209–218.
29. Campbell, JE, RP Thompson, J Dean and TW Clyne, Comparison between stress–strain plots obtained from indentation plastometry, based on residual indent profiles, and from uniaxial testing. *Acta Materialia*, 2019. **168**: 87–99.

30. Cottrell, AH and BA Bilby, Dislocation theory of yielding and strain ageing of iron. *Proceedings of the Physical Society of London Section A*, 1949. **62**(349): 49–62.
31. Brindley, BJ, RW Honeycombe and DJ Corderoy, Yield points and Luders bands in single crystals of copper-base alloys. *Acta Metallurgica*, 1962. **10**(Nov): 1043–1050.
32. Neuhauser, H and A Hampel, Observation of Luders bands in single crystals. *Scripta Metallurgica et Materialia*, 1993. **29**(9): 1151–1157.
33. Lloyd, DJ and LR Morris, Luders band deformation in a fine-grained aluminium alloy. *Acta Metallurgica*, 1977. **25**(8): 857–861.
34. Balasubramanian, N, JCM Li and M Gensamer, Plastic deformation and Luders band propagation in alpha brass. *Materials Science and Engineering*, 1974. **14**(1): 37–45.
35. Kyriakides, S and JE Miller, On the propagation of Luders bands in steel strips. *Journal of Applied Mechanics: Transactions of the ASME*, 2000. **67**(4): 645–654.
36. Gorbatenko, VV, VI Danilov and LB Zuev, Plastic flow instability: Chernov–Luders bands and the Portevin–Le Chatelier effect. *Technical Physics*, 2017. **62**(3): 395–400.
37. Khotinov, VA, ON Polukhina, DI Vichuzhan, GV Schapov and VM Farber, Study of Luders deformation in ultrafine low-carbon steel by the digital image correlation technique. *Letters on Materials*, 2019. **9**(3): 328–333.
38. Zuev, LB, VV Gorbatenko and VI Danilov, Chernov–Luders bands and the Portevin–Le Chatelier effect as plastic flow instabilities. *Russian Metallurgy*, 2017(4): 231–236.
39. Wang, XG, L Wang and MX Huang, In-situ evaluation of Luders band associated with martensitic transformation in a medium Mn transformation-induced plasticity steel. *Materials Science and Engineering A: Structural Materials Properties Microstructure and Processing*, 2016. **674**: 59–63.
40. Jafarian, H, Characteristics of nano/ultrafine-grained austenitic trip steel fabricated by accumulative roll bonding and subsequent annealing. *Materials Characterization*, 2016. **114**: 88–96.
41. Cai, MH, WJ Zhu, N Stanford, LB Pan, Q Chao and PD Hodgson, Dependence of deformation behavior on grain size and strain rate in an ultrahigh strength-ductile Mn-based trip alloy. *Materials Science and Engineering A: Structural Materials Properties Microstructure and Processing*, 2016. **653**: 35–42.
42. Louche, H and A Chrysochoos, Thermal and dissipative effects accompanying Luders band propagation. *Materials Science and Engineering A: Structural Materials Properties Microstructure and Processing*, 2001. **307**(1–2): 15–22.
43. Murav'ev, TV and LB Zuev, Acoustic emission during the development of a Luders band in a low-carbon steel. *Technical Physics*, 2008. **53**(8): 1094–1098.
44. Hauser, JJ and KA Jackson, Effect of grip constraints on the tensile deformation of FCC single crystals. *Acta Metallurgica*, 1961. **9**(1): 1–13.
45. Kim, JY and JR Greer, Tensile and compressive behavior of gold and molybdenum single crystals at the nano-scale. *Acta Materialia*, 2009. **57**(17): 5245–5253.
46. Sowerby, R and W Johnson, Review of texture and anisotropy in relation to metal forming. *Materials Science and Engineering*, 1975. **20**(2): 101–111.
47. Kalidindi, SR, Modeling anisotropic strain hardening and deformation textures in low stacking fault energy FCC metals. *International Journal of Plasticity*, 2001. **17**(6): 837–860.
48. Dawson, PR, SR MacEwen and PD Wu, Advances in sheet metal forming analyses: dealing with mechanical anisotropy from crystallographic texture. *International Materials Reviews*, 2003. **48**(2): 86–122.

49. Wenk, HR and P Van Houtte, Texture and anisotropy. *Reports on Progress in Physics*, 2004. **67**(8): 1367–1428.
50. Tucker, GEG, Texture and earing in deep drawing of aluminium. *Acta Metallurgica*, 1961. **9**(4): 275–286.
51. Zhao, Z, W Mao, F Roters and D Raabe, A texture optimization study for minimum earing in aluminium by use of a texture component crystal plasticity finite element method. *Acta Materialia*, 2004. **52**(4): 1003–1012.
52. Raabe, D, Y Wang and F Roters, Crystal plasticity simulation study on the influence of texture on earing in steel. *Computational Materials Science*, 2005. **34**(3): 221–234.
53. Tiernan, P and A Hannon, Design optimisation of biaxial tensile test specimen using finite element analysis. *International Journal of Material Forming*, 2014. **7**(1): 117–123.
54. Xiao, R, A review of cruciform biaxial tensile testing of sheet metals. *Experimental Techniques*, 2019. **43**(5): 501–520.
55. Teaca, M, I Charpentier, M Martiny and G Ferron, Identification of sheet metal plastic anisotropy using heterogeneous biaxial tensile tests. *International Journal of Mechanical Sciences*, 2010. **52**(4): 572–580.
56. Nicholas, T, Tensile testing of materials at high rates of strain. *Experimental Mechanics*, 1981. **21**(5): 177–185.
57. Ellwood, S, LJ Griffiths and DJ Parry, A tensile technique for materials testing at high strain rates. *Journal of Physics E: Scientific Instruments*, 1982. **15**(11): 1169–1172.
58. Smerd, R, S Winkler, C Salisbury, M Worswick, D Lloyd and M Finn, High strain rate tensile testing of automotive aluminum alloy sheet. *International Journal of Impact Engineering*, 2005. **32**(1–4): 541–560.
59. Korhonen, AS and HJ Kleemola, Effects of strain rate and deformation heating in tensile testing. *Metallurgical Transactions A: Physical Metallurgy and Materials Science*, 1978. **9**(7): 979–986.

6 Compressive Testing

Testing in (uniaxial) compression is sometimes an attractive alternative to tensile testing. Specimens can be simpler in shape and smaller, since there is no gripping requirement. The key question is whether corresponding information can be obtained. In general, it can, but there is sometimes a perception that at least some materials behave differently under compression – i.e. that there is tensile-compressive asymmetry in their response. In fact, this is largely a myth: at least in the majority of cases, the underlying plasticity response is symmetrical (and indeed the von Mises (deviatoric) stress, which is normally taken to be the determinant of the response, is identical in the two cases). However, there are important caveats to append to this statement. For example, if the material response is indeed dependent on the hydrostatic component of the stress, as it might be for porous materials and for those in which a phase transformation occurs during loading, then asymmetry is possible. Also, while the underlying plasticity response is usually the same, the compressive stress–strain curve is often affected by friction between sample and platen (leading to barreling). Conversely, the necking that is likely to affect the tensile curve cannot occur in compression, although some kind of buckling or shearing instability is possible. It's also important to distinguish the concept of tension/compression asymmetry from that of the Bauschinger effect (a sample pre-loaded in tension exhibiting a different response if then loaded in compression).

6.1 Test Configuration

6.1.1 Sample Geometry, Strain Measurement and Lubrication Issues

The most common sample shape is a simple cylinder. Of course, other sectional shapes, such as a square or rectangle, are possible, but it is certainly conventional for it to be prismatic (uniform in section along the loading direction). Unlike the tensile test, this is normally the case along the complete length of the sample, although there is still the option of having a gauge length (where the strain is monitored) that is less than the whole sample length.

However, there is often a difficulty with compressive testing concerning the total sample length. As outlined in §6.1.2, the aspect ratio (length/diameter) has to be kept low if the danger of buckling is to be avoided. A value of around 2 is often an upper limit and commonly the ratio is close to unity. Even if the whole sample length is taken as the gauge length, which is commonly done, this tends to have the effect of making the distance over which the axial strain is to be measured relatively small. Of course, there is the option of raising the diameter, but this will increase the load requirements. As noted in §5.1.3, if the diameter is much more than about 10 mm, then, at least for strong materials, the load needed to cause plastic deformation might start to get relatively large (>~200 kN).

Accurate measurement of strain can thus present a challenge, particularly since the presence of the platens (which are often much larger in diameter than the sample) can make it difficult to fit clip gauges. There are various other options, such as using an LVDT, an eddy current gauge or a scanning laser extensometer to measure the distance between the platens, although there may be limitations on the accuracy due to the short sample length and there may be complications such as the effects of squeezing out of a lubricant introduced between sample and platen – see below.

One of the main issues for compression testing is the effect of *friction* between sample and platens. The simplest scenario is that of a *frictionless* interface, such that, as straining continues, the sample retains a prismatic shape, simply becoming shorter and wider. This requires **unhindered sliding** at the interface and ensures that the stress field remains uniform (and uniaxial). There is still the issue of the true stress (and strain) diverging from the nominal values. This occurs in a similar way to that during tension, except that the true stress starts to drop below the nominal value (and the true strain gets larger than the nominal strain – i.e. the reverse of what happens in tension). The other extreme case is when there is **no interfacial sliding** (and pronounced "*barreling*" occurs as the test progresses). This is occasionally referred to as "*sticking friction*," although there is scope for confusion in this nomenclature, since it is sometimes taken to mean that sliding can occur, but only with a shear stress equal to the shear yield stress of the sample. The case of no sliding is at least a well-defined condition that can readily be incorporated into models.

In reality the condition at the interface normally lies between these extremes, and is characterized by the value of a friction coefficient – see §6.2.2. Depending on its value, and the applied load, some interfacial sliding is likely to occur during the test, but it will be at least partially inhibited and there will be a degree of barreling. From a practical point of view, it is common to **lubricate** the interface in an attempt to minimize the effects of friction. Various lubricants are available [1–3], with those based on $MoSi_2$ being a common choice. However, it may be noted that, as a consequence of the high contact pressure during the test, there is often a tendency for much of the lubricant to be squeezed out. This increases the friction and may make it variable during the test, as well as possibly contributing to the measured strain (if obtained via the distance between the platens).

6.1.2 Buckling Instabilities

As mentioned above, the aspect ratio of the sample needs to be kept relatively small if the danger of buckling is to be avoided. In fact, elastic (or Euler) buckling is unlikely.

For compression testing, in which the ends cannot normally rotate or translate, the **Euler buckling stress formula** (for a cylinder of length L and diameter D) may be written

$$\sigma_b = \frac{\pi^2 E}{4(L/D)^2} \tag{6.1}$$

Taking E to be 200 GPa and an upper limit for the required stress (to complete the test) to be 3 GPa, this formula suggests that it's only necessary to keep (L/D) below about 10 to avoid buckling. However, this is misleading. Buckling tends to occur more readily after the onset of plasticity, when the sample is deforming as if it were effectively less stiff. Moreover, the instability may not occur with the classical buckling geometry, but may be more one of a lateral shearing of one end relative to the other. This type of process is much harder to analyze, but it can certainly occur with a much lower aspect ratio than that needed for elastic buckling. This is the kind of issue that can often be resolved only by experiment, but in general it's advisable to keep the ratio below about 2, and possibly to avoid exceeding unity. Detailed information about buckling (elastic and plastic), including the behavior of hollow tubes under compression, is provided in §10.2.

6.1.3 Sample Size and Micropillar Compression

As for tensile testing (§5.1.3), there are issues related to sample size, and particularly to small samples. When the sample section is very small, even if only in one direction, there are concerns about whether the constraint conditions, and the number of grains across the section, are likely to differ significantly from those of a "bulk" sample, perhaps resulting in deduced (plasticity) properties that are different from those of the bulk. Data such as those in Fig. 5.4 confirm that this can occur.

Under compression, similar concerns apply, although the need to avoid buckling tends to militate against use of samples with small transverse dimensions (even in only one direction). Furthermore, there are difficulties in measuring the strain for samples with very short gauge lengths, such as for compressive testing of thin sheet in the through-thickness direction. Nevertheless, there has been considerable interest [4–7] in the compressive testing of very short samples (with suitably low aspect ratios) – i.e. in so-called "*micropillar compression*" – see §9.5. The loading is usually carried out using a "nanoindenter" system, via a cylindrical punch. These samples, which are commonly produced by etching or ion beam milling techniques, typically have dimensions of the order of a few microns. They are thus usually single crystals, or contain relatively few grains, and so are unlikely to exhibit closely similar behavior to that of the (bulk) samples from which they have been milled. There are also concerns about the constraint conditions, which are different at the loading end from those at the base, where the cylinder is embedded (continuous with the substrate from which it has been milled). Finally, it has become clear that the milling operation can leave embedded ions and generally affect the structure (and properties). Details are provided in §9.5.

It seems clear that the technique cannot be used to obtain reliable (plasticity) properties that reflect those of the material from which they have been milled. The usage, which has been quite extensive, has tended to focus more on the basics of single crystal deformation, with it being possible to observe dynamically the formation of slip bands etc. – the testing is commonly carried out within a scanning electron microscope (SEM) (and indeed it is often difficult to load the pillar successfully without being able to view the sample effectively during positioning).

6.2 Compressive Stress–Strain Curves

6.2.1 Nominal and True Stress–Strain Plots

As for tensile testing, it's a simple matter to convert an experimental (nominal) stress–strain curve to one expressed as true values (using Eqns. (3.10) and (3.11), which are based on the assumption of uniform stress and strain fields). However, while this assumption is expected to hold during tensile testing up to the onset of necking, it is likely to be invalid from the start in compressive testing (unless there is no interfacial friction). Nevertheless, it's quite common to ignore this effect.

Representative stress–strain curves [8] are shown in Fig. 6.1, for Cu samples in two different conditions – as-received and annealed. (Extruded Cu bars in these two conditions are compared at several places in the book, but it's important to note that the true stress–strain curves for each of them are not identical in all cases, since they are not all from the same stock material and the curve tends to be very sensitive to details of purity and processing – see §4.3.3.) The comparison shown in Fig. 6.1(c) is between the true plots obtained from the tensile and compressive tests (using Eqns. (3.10) and (3.11)). It should first be understood that the tensile-derived plots are not expected to be valid beyond the necking point. Moreover, the compressive-derived plots are likely to be influenced by friction (and hence to be invalid) from the start, although the effect may be small initially (particularly if there was effective lubrication).

It can be seen that there is certainly a large measure of agreement. The discrepancy is probably attributable to the squeezing out of lubricant during the test, which raises the apparent strain (to a degree that increases as the stress is raised). This could have been eliminated if the strain had been measured directly on the sample, rather than between the platens, although, as outlined in §6.1.1, this is not easy to do. In fact, the agreement with the tensile curves is better if the unlubricated compressive plots are used. In this case there was no squeezing out of lubricant, although the error arising from frictional effects would be expected to be greater than for the

6.2 Compressive Stress–Strain Curves

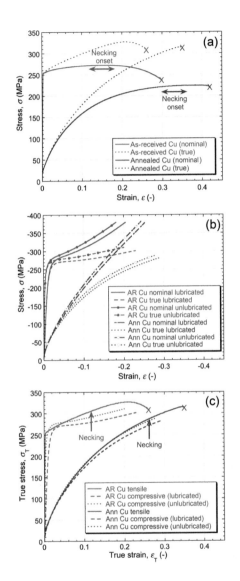

Fig. 6.1 Stress–strain data [8] (nominal and true curves), for as-received (AR) and annealed (Ann) Cu, in (a) tension, (b) compression (with and without lubrication) and (c) comparison for the true curves.

lubricated case. These data suggest that such effects were relatively small in this case, at least over the initial range prior to necking in the tensile tests. These data demonstrate that, in general, the same underlying (plasticity) characteristics are obtained from tensile and compressive testing, although they also highlight that certain effects need to be taken into account in order to avoid the conclusion that there is some kind of (tensile/compressive) asymmetry. Genuine asymmetry of this type is possible, but in fact is relatively rare – see §6.3.1.

6.2.2 FEM Modeling, Sample/Platen Friction and Barreling Effects

As outlined above, it is difficult to eliminate friction at the sample/platen interface. It is easy to recognize that such interfacial friction has played a role, since it leads to the development of a "barrel" shape (for a sample that is initially cylindrical). Also, by measuring the diameter at the top and bottom of the sample at the end of the test, and comparing it with the initial sample diameter, it's possible to check whether there has in fact been any interfacial sliding.

On a more quantitative level, if the (true) stress–strain relationship of the material is known, then FEM simulation of the compression test can be used to characterize the friction conditions, via an inverse (iterative) modeling sequence [9–14]. This characterization most commonly takes the form of a value for the *coefficient of friction*, μ, which is the ratio of the shear stress necessary for interfacial sliding to the normal (compressive) stress on the interface – see Eqn. (8.1). Some results illustrating the approach are shown in Fig. 6.2, taken from the work of Wang et al. [13]. This shows strain fields after an axial compression of about −50% (nominal, which corresponds, using Eqn. (3.11), to a true strain of about −70%). For the frictionless (uniform strain field) case, the von Mises strain is actually slightly higher

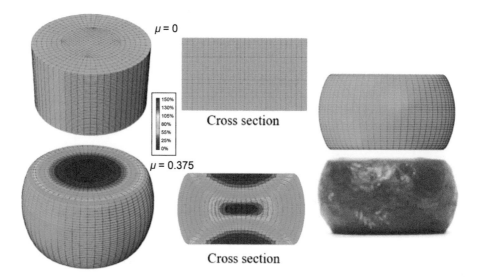

Fig. 6.2 Fields of equivalent plastic strain for a low C steel [13], after uniaxial compression with and without a finite coefficient of friction, and a comparison between a modeled and experimentally observed barreling shape.

than this, since it incorporates the effect of the transverse strains – see §3.1.1. It is clear in this figure that frictional effects can cause the strain field to become highly inhomogeneous.

An indication is given in Fig. 6.2 of how the barreling shape is used as an experimental outcome to guide evaluation of the friction coefficient (once the true stress–strain relationship of the material has been obtained). In the work concerned, which was carried out at high temperature (~1000–1200 °C), over a range of strain rate, the best-fit value of μ was found to be about 0.375. This is a relatively high value, although that is often the case for processing carried out at high temperature.

Of course, it's also possible to use FEM simulation to predict stress–strain curves in compression, taking account of frictional effects, and using an appropriate true stress–strain relationship. (This can also be done for tension, taking account of necking effects, as described in §5.3.3.) A comparison [8] is shown in Fig. 6.3 between predicted and experimental stress–strain curves, for two materials, with and without lubrication. Of course, friction is relevant, but it's also worth noting that there is often an initial "***bedding down***" effect that presents a complication in terms of comparing model predictions with experiment. There is no real prospect of incorporating the latter into a model, although this can be done for the former (via a coefficient of friction, μ).

What was done here was to use the two best-fit sets of parameter values (obtained from tensile comparisons [8]). There is information available in the literature about likely values of μ under different conditions [12, 14–17]. A value in the approximate range 0.2–0.3 has often been found appropriate for unlubricated compression, although clearly there may be a dependence on surface finish, materials etc. During lubricated compression testing, there tends to be more variation, but a value of the order of 0.05–0.1 might be considered typical with good lubrication. Accepting that accurate estimation of μ is difficult, and also that it may change during the process, values of 0.1 and 0.3 were used in the work of Fig. 6.3, designed to correspond to the lubricated and unlubricated cases. It can be seen that, for both materials, there is a fairly good level of agreement between experiment and prediction. The differences between the high and low μ predictions are certainly similar to those of the two experimental conditions. There is clearly an error associated with the "bedding down" process, leading to larger strains over the complete range for the experimental plots. Accepting this, however, and recognizing that, for the most accurate comparisons, it is probably best not to use compressive uniaxial data, the level of consistency is reasonably good (confirming that the plasticity characteristics are being well captured by these two parameter sets for these two materials).

The initial ("elastic") slope is lower when lubricant is present, although this is only noticeable for the AR-Cu (with the higher yield stress). This is presumably due to the "bedding down," which is more pronounced in the presence of lubricant – probably as a result of it being progressively squeezed out of the interface during the early part of the test. However, even the unlubricated case does show some evidence of such an

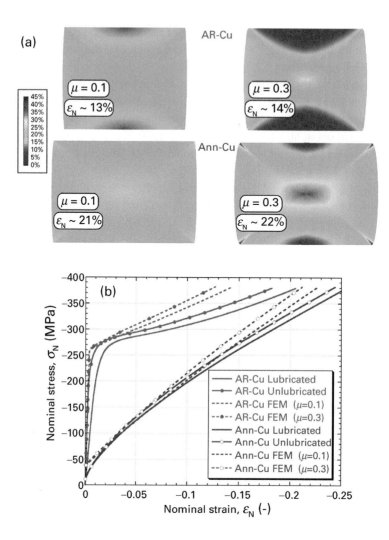

Fig. 6.3 Comparison [8] between FEM and experiment during compressive loading of two materials, showing (a) von Mises plastic strain fields (at the nominal strains shown), for two values of μ, and (b) experimental (with and without lubrication) and predicted (nominal) stress–strain curves.

effect, which can arise if the surfaces coming into contact are not perfectly flat and parallel. Bedding down effects could be avoided by using a clip gauge, but this is difficult for compression testing. Also, the absence of lubricant does lead to a (small, but noticeable) progressive increase in the stress needed to continue straining, compared with the lubricated case. This is more noticeable for the Ann-Cu. It is a consequence of μ being higher when there is no lubricant. It should, however, be noted that, in reality, the value of μ may change during a test.

A general conclusion about compressive testing is that, while it can be used to obtain reliable information about the underlying plasticity response of the material, it is more susceptible to the effects of variables that are difficult to pin down accurately than is the case with tensile testing. Of course, tensile testing has various disadvantages and there are strong arguments in favor of indentation plastometry, which is even easier and more versatile than compressive testing, and is potentially more reliable and accurate (see Chapter 8).

6.3 Tension/Compression Asymmetry and the Bauschinger Effect

6.3.1 Tension/Compression Asymmetry

There is quite frequent reference in the literature to "tension/compression asymmetry," meaning a difference between the inherent (plasticity) responses of a material when subjected to (uniaxial) compression and tension. It may first be noted that this implies a dependence on the hydrostatic component of the stress state (since the deviatoric, i.e. von Mises, stress is the same in both cases – see §3.1.1). If true, this would be potentially a cause for serious concern, since virtually all FEM modeling (carried out globally on a massive scale) is based on using constitutive laws involving only the von Mises stress.

Of course, that is not a basis on which to dismiss these claims, which should naturally be scrutinized on their merits, both in terms of experimental evidence and from the theoretical point of view. It should also be emphasized that, for certain types of material, a dependence on the hydrostatic stress IS expected. The most obvious of these is *porous materials*, which can for these purposes be regarded as those with porosity levels above a few %. Pores are likely to become closed when the hydrostatic stress is negative (compressive) and opened up when it is tensile. This will certainly lead to different (plasticity) responses in tension and compression, although, since it might be expected to effectively reduce the hardness in both cases, the expected direction of the asymmetry is not immediately clear. It is in any event recognized in the literature [18–20] that asymmetry can arise in such materials.

A little less obvious is that asymmetry might also be expected when the plasticity is accompanied by a phase transformation (with an associated volume change). If the volume decreases during the phase change, then it will be promoted by a compressive hydrostatic stress and vice versa. A key issue here is the magnitude of the volume change, which might be very small (in which case the effect is expected to be weak). However, phase transformations can certainly be accompanied by quite significant volume changes. Those stimulated by mechanical loading are likely to be *martensitic* (*diffusionless*), since these can occur quickly. They are described in §4.4.2, where attention is mostly focussed on those in *superelastic* (and *shape memory*) alloys. Indeed, a proportion of the reports of tensile/compressive asymmetry do relate to such materials [21–26].

In fact, shape memory alloys represent a very small proportion of metal utilization and in any event their mechanical characteristics are complex (§3.2.3 and §4.4.2), such that any tension/compression asymmetry would introduce only a minor modification to them. However, there are other types of system in which mechanically induced martensitic transformations can occur, particularly certain types of steel, and these are much more widely used and industrially significant than shape memory alloys. Indeed, the class often referred to as **TRIP** (*transformation-induced plasticity*) **steels** [27–30] has the mechanical stimulation of martensitic transformations as a basic characteristic. This type of deformation also often occurs in the so-called "*dual phase*" steels. These are high strength steels with good formability, usually having a ferrite-based microstructure containing relatively high levels of martensite. The (soft) ferrite gives a relatively low yield stress, but the (hard) martensite, and potentially the formation of further martensite during the loading, confers a high work hardening rate, so the UTS of such materials is high.

Similar hardening as the load is increased, also largely due to stimulation of phase transformations, also occurs in **Hadfield's manganese steel** ("***Mangalloy***"). Figure 6.4 shows true stress–true strain plots [31], obtained from tensile and compressive tests, for such a steel. The high work hardening rate, raising the flow stress by about 1 GPa over a strain range of about 25% (i.e. a more or less linear work hardening rate of about 4 GPa), is immediately apparent. It also seems clear that there is tensile/compressive asymmetry, with the material being harder in compression than tension. Caveats should, however, be appended to this. Conversion from nominal to true curves was apparently carried out using the analytical relationships. For the tensile plots, which were unaffected by necking, this should be accurate. For the compressive ones, however, friction could have had an effect, as described in §6.2.2. There is no

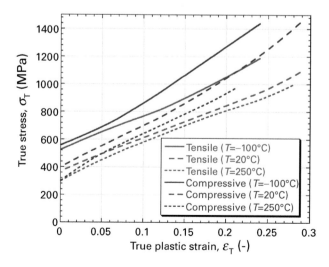

Fig. 6.4 True stress–strain plots for Hadfield's manganese steel, obtained from uniaxial testing in tension and compression at different temperatures [31].

reference in the paper to lubrication or assessment of friction and it is possible that the observation of higher flow stress values in compression is largely attributable to frictional effects. Nevertheless, it is possible that at least some genuine asymmetry was arising from the contribution of phase transformations to the straining, although any such effect was probably small.

However, it seems very likely that many of the reports of asymmetry, which certainly cover many systems in which there is neither porosity nor stimulation of martensitic phase transformations during loading, actually arose entirely from imperfect conversion of the raw data to true stress–strain curves. As outlined above, it is not a simple matter to carry out these conversions to high accuracy, particularly for compressive testing, and there are very few reports in which a large and unambiguous asymmetry has been found experimentally. It should also be mentioned that there have been various attempts to provide a theoretical basis for tensile/compressive asymmetry. Many of these involve arguments about dislocation mobility and/or twinning, often invoking crystallographic texture in some way to explain the asymmetry [32–36]. It is, of course, possible that individual explanations may have some validity, but in general it appears unlikely that these mechanisms would lead to strong asymmetries of any sort. In some cases, there may be some confusion with the Bauschinger effect, a well-established phenomenon that is described in the next section. As a generalization, neglect of the hydrostatic component of the stress state in an analysis of plasticity characteristics is usually an acceptable assumption.

6.3.2 The Bauschinger Effect

This effect was first identified in 1881 by Johann Bauschinger. He observed that, when a sample was deformed plastically in tension, and then tested in compression, the yield stress was lower than it had been in tension. This is, of course, a different effect from that of a tensile/compressive asymmetry (for testing of different samples of the same material). It suggests that something has happened during the first test that has affected its response during the second test (and, in practice, the effect is often investigated via cyclic tests, with repeated reversal of the sense of the loading). It has been the subject of extensive investigation [37–41], which has revealed that it occurs in single crystals, as well as polycrystalline samples. A schematic plot [42], and some experimental data [38], are shown in Fig. 6.5. The phenomenon can be regarded as broader than just a dependence of the flow stress on prior loading in the reverse direction, since it raises the possibility of any plasticity characteristics being dependent on prior strain history, potentially creating *anisotropy*.

There are several different ways in which the effect has been explained, but the main proposed mechanisms are based either on the generation of *residual stresses* or on *dislocation mobility*. The concept of a *"back-stress"* is central to both approaches. The idea of the initial (tensile) loading creating residual stresses that facilitate yielding under the reversed (compressive) load has been popular. However, a simple uniaxial test, with uniform stress and strain fields, should not create any residual stresses – these arise only when there is some kind of differential straining (and they must *force*

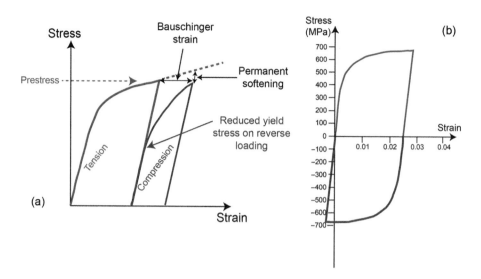

Fig. 6.5 (a) Schematic [42] and (b) experimental [38] (for an X-80 grade steel) stress–strain plots, illustrating the Bauschinger effect.

balance to zero when integrated over the sample – see §3.3.1). If the effect is to be explained in terms of residual stresses, then they must be local ones (that facilitate reverse plasticity). Explanations also focus on dislocation motion becoming inhibited (in the "forward" direction) during plastic deformation, but then being easier when the loading was reversed. Initially, it was envisaged that this inhibition was largely in the form of "pile-ups" at grain boundaries, but observation of the effect in single crystals made it clear that grain boundaries were not essential. Nevertheless, the most plausible explanation is that dislocations have encountered various kinds of obstacles, tangles and so on (when moving in one direction) and that it is easier, at least initially, for them to start moving in the reverse direction. It's also possible to explain this in terms of (residual) local stress fields.

Of course, such a situation (i.e. dislocations being "pinned" in some way, and inhibited from continuing to move in the directions being promoted by the applied load) could exist in various samples, particularly those in a "work-hardened" state – for example, after being extruded or rolled etc. Equivalently, this situation could be considered in terms of the presence of residual stresses. Such states could even give rise to a tensile/compressive asymmetry. It can, however, be argued that this is rather unlikely, since metal-working processes of this type tend to create rather complex residual stress fields that would balance out over the sample as a whole in terms of their effect during a uniaxial test. On the other hand, for hardness testing or indentation plastometry, in which only small parts of a sample are being mechanically interrogated, the residual stress in the part concerned may influence the outcome of the test (with potential scope for measuring the residual stress in the region concerned – see §8.4.3).

6.4 The Ring Compression Test

The ring compression test [43–47] is specifically aimed at characterizing the interfacial friction (for given materials, surface finishes, temperatures, lubricant type etc.). It differs from a standard compression test in having a central hole (in a cylindrical sample). Since material can flow so as to reduce the diameter of the inner hole, as well as expanding the outer diameter, the response is more sensitive to the frictional conditions. A typical ratio of outer diameter: inner diameter: height is 6:3:2 or 4:2:1. The procedure usually involves iterative FEM simulation of the process (with best-fit plasticity characteristics), so as to obtain an optimized value of μ. The target outcome could involve barreling shapes, although in its original form it was simply the relative reduction in inner diameter, as a function of the height reduction [43], with no account taken of the detailed plasticity response of the sample. In fact, this diameter can actually increase if the coefficient of friction is very low, but it's more common for it to be reduced.

The data [47] shown in Fig. 6.6, which are from a detailed investigation based on FEM simulation, give an indication of the factors affecting the value of μ for dry friction (i.e. no lubrication). They refer to samples of Pb compressed between stainless steel platens, with the roughness of the platen surfaces and the thickness of the oxide film on the samples being varied. It can be seen that the surface roughness has an effect, with the friction coefficient rising from about 0.2 to around 0.6 as the roughness is raised from a low value to about 0.3 µm. The oxide film thickness, conversely, has relatively little effect. This behavior can be rationalized [47] in terms of deformation mechanisms at the interface. Of course, the presence of lubricant can change the behavior substantially. It's also worth noting that Pb is very soft, so even a situation in which it is essentially adhering to the platens ("sticking friction"), doesn't lead to very high interfacial shear stresses.

In general, while the ring compression test is potentially informative about frictional conditions, and can often be related to the response of a component during forging, it is not commonly used for standard compression testing aimed at obtaining

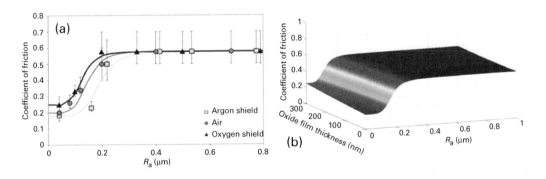

Fig. 6.6 Measured values [47] of the coefficient of friction, obtained using the ring compression test with Pb samples and stainless steel platens, plotted against: (a) surface roughness of the platens and (b) surface roughness and the thickness of the oxide film on the samples.

the plasticity characteristics of material. This is partly because it requires the sample to have a fairly large diameter, and for it to be possible to drill a hole in it. However, if the material is available in dimensions suitable for this, then there is an argument for using this test in preference to, or possibly in parallel with, a conventional test, since accurate evaluation of the effect of friction is essential if reliable (true stress–true strain) plasticity characteristics are to be obtained. In any event, iterative FEM simulation of the test is required.

References

1. Mizuno, T and M Okamoto, Effects of lubricant viscosity at pressure and sliding velocity on lubricating conditions in the compression-friction test on sheet metals. *Journal of Lubrication Technology: Transactions of the ASME*, 1982. **104**(1): 53–59.
2. Li, LX, DS Peng, JA Liu and ZQ Liu, An experiment study of the lubrication behavior of graphite in hot compression tests of Ti-6Al-4V Alloy. *Journal of Materials Processing Technology*, 2001. **112**(1): 1–5.
3. Li, P, CR Siviour and N Petrinic, The effect of strain rate, specimen geometry and lubrication on responses of aluminium AA2024 in uniaxial compression experiments. *Experimental Mechanics*, 2009. **49**(4): 587–593.
4. Frick, CP, BG Clark, S Orso, AS Schneider and E Arzt, Size effect on strength and strain hardening of small-scale 111 nickel compression pillars. *Materials Science and Engineering A: Structural Materials Properties Microstructure and Processing*, 2008. **489**(1–2): 319–329.
5. Fei, HY, A Abraham, N Chawla and HQ Jiang, Evaluation of micro-pillar compression tests for accurate determination of elastic–plastic constitutive relations. *Journal of Applied Mechanics: Transactions of the ASME*, 2012. **79**(6).
6. Chen, R, S Sandlobes, C Zehnder, XQ Zeng, S Korte-Kerzel and D Raabe, Deformation mechanisms, activated slip systems and critical resolved shear stresses in an Mg-LPSO alloy studied by micro-pillar compression. *Materials & Design*, 2018. **154**: 203–216.
7. Ying, SQ, LF Ma, T Sui, C Papadaki, E Salvati, LR Brandt, HJ Zhang and AM Korsunsky, Nanoscale origins of the size effect in the compression response of single crystal Ni-base superalloy micro-pillars. *Materials*, 2018. **11**(4).
8. Campbell, JE, RP Thompson, J Dean and TW Clyne, Comparison between stress–strain plots obtained from indentation plastometry, based on residual indent profiles, and from uniaxial testing. *Acta Materialia*, 2019. **168**: 87–99.
9. Li, YP, E Onodera and A Chiba, Friction coefficient in hot compression of cylindrical sample. *Materials Transactions*, 2010. **51**(7): 1210–1215.
10. Yao, ZH, DQ Mei, H Shen and ZC Chen, A friction evaluation method based on barrel compression test. *Tribology Letters*, 2013. **51**(3): 525–535.
11. Zhou, J, P He, JF Yu, LJ Lee, LG Shen and AY Yi, Investigation on the friction coefficient between graphene-coated silicon and glass using barrel compression test. *Journal of Vacuum Science & Technology B*, 2015. **33**(3).
12. Duran, D and C Karadogan, Determination of Coulomb's friction coefficient directly from cylinder compression tests. *Strojniski Vestnik – Journal of Mechanical Engineering*, 2016. **62**(4): 243–251.

13. Wang, X, H Li, K Chandrashekhara, SA Rummel, S Lekakh, DC Van Aken and RJ O'Malley, Inverse finite element modeling of the barreling effect on experimental stress–strain curve for high temperature steel compression test. *Journal of Materials Processing Technology*, 2017. **243**: 465–473.
14. Torrente, G, Numerical and experimental studies of compression-tested copper: proposal for a new friction correction. *Materials Research – Ibero – American Journal of Materials*, 2018. **21**(4).
15. Fardi, M, R Abraham, PD Hodgson and S Khoddam, A new horizon for barreling compression test: exponential profile modeling. *Advanced Engineering Materials*, 2017. **19**(11).
16. Bol, M, R Kruse and AE Ehret, On a staggered iFEM approach to account for friction in compression testing of soft materials. *Journal of the Mechanical Behavior of Biomedical Materials*, 2013. **27**: 204–213.
17. Fan, XG, YD Dong, H Yang, PF Gao and M Zhan, Friction assessment in uniaxial compression test: a new evaluation method based on local bulge profile. *Journal of Materials Processing Technology*, 2017. **243**: 282–290.
18. Deng, X, GB Piotrowski, JJ Williams and N Chawla, Effect of porosity and tension–compression asymmetry on the Bauschinger effect in porous sintered steels. *International Journal of Fatigue*, 2005. **27**(10–12): 1233–1243.
19. Stewart, JB and O Cazacu, Analytical yield criterion for an anisotropic material containing spherical voids and exhibiting tension–compression asymmetry. *International Journal of Solids and Structures*, 2011. **48**(2): 357–373.
20. Alves, JL, MC Oliveira, LF Menezes and O Cazacu, The role of tension–compression asymmetry of the plastic flow on ductility and damage accumulation of porous polycrystals. *Ciencia & Tecnologia dos Materiais*, 2017. **29**(1): E234–E238.
21. Gall, K and H Sehitoglu, The role of texture in tension–compression asymmetry in polycrystalline NiTi. *International Journal of Plasticity*, 1999. **15**: 69–92.
22. Gall, K, H Sehitoglu, YI Chumlyakov and IV Kireeva, Tension–compression asymmetry of the stress–strain response in aged single crystal and polycrystalline NiTi. *Acta Materialia*, 1999. **47**: 1203–1217.
23. Adharapurapu, RR, F Jiang, KS Vecchio and GT Gray III, Response of NiTi shape memory alloy at high strain rate: a systematic investigation of temperature effects on tension–compression asymmetry. *Acta Materialia*, 2006. **54**(17): 4609–4620.
24. Grolleau, V, H Louche, V Delobelle, A Penin, G Rio, Y Liu and D Favier, assessment of tension–compression asymmetry of NiTi using circular bulge testing of thin plates. *Scripta Materialia*, 2011. **65**(4): 347–350.
25. Ma, J, B Kockar, A Evirgen, I Karaman, ZP Luo and YI Chumlyakov, Shape memory behavior and tension–compression asymmetry of a FeNiCoAlTa single-crystalline shape memory alloy. *Acta Materialia*, 2012. **60**(5): 2186–2195.
26. Bucsek, AN, HM Paranjape and AP Stebner, Myths and truths of nitinol mechanics: elasticity and tension–compression asymmetry. *Shape Memory and Superelasticity*, 2016. **2**(3): 264–271.
27. Sugimoto, K, N Usui, M Kobayashi and S Hashimoto, Effects of volume fraction and stability of retained austenite on ductility of trip-aided dual phase steels. *ISIJ International*, 1992. **32**(12): 1311–1318.
28. De Cooman, BC, Structure-properties relationship in trip steels containing carbide-free bainite. *Current Opinion in Solid State & Materials Science*, 2004. **8**(3–4): 285–303.

29. Kim, H, J Park, Y Ha, W Kim, SS Sohn, HS Kim, BJ Lee, NJ Kim and S Lee, Dynamic tension–compression asymmetry of martensitic transformation in austenitic Fe-(0.4,1.0)C-18Mn steels for cryogenic applications. *Acta Materialia*, 2015. **96**: 37–46.
30. Joo, G and H Huh, Rate-dependent isotropic–kinematic hardening model in tension–compression of TRIP and TWIP steel sheets. *International Journal of Mechanical Sciences*, 2018. **146**: 432–444.
31. Adler, PH, GB Olson and WS Owen, Strain hardening of Hadfield manganese steel. *Metallurgical Transactions A: Physical Metallurgy and Materials Science*, 1986. **17**(10): 1725–1737.
32. Cheng, S, JA Spencer and WW Milligan, Strength and tension/compression asymmetry in nanostructured and ultrafine-grain metals. *Acta Materialia*, 2003. **51**(15): 4505–4518.
33. Luo, H, L Shaw, LC Zhang and D Miracle, On tension/compression asymmetry of an extruded nanocrystalline Al-Fe-Cr-Ti alloy. *Materials Science and Engineering: A Structural Materials: Properties, Microstructure and Processing*, 2005. **409**(1–2): 249–256.
34. Yapici, GG, IJ Beyerlein, I Karaman and CN Tome, Tension–compression asymmetry in severely deformed pure copper. *Acta Materialia*, 2007. **55**(14): 4603–4613.
35. Lv, CL, TM Liu, DJ Liu, S Jiang and W Zeng, Effect of heat treatment on tension–compression yield asymmetry of AZ80 magnesium alloy. *Materials & Design*, 2012. **33**: 529–533.
36. Park, SH, JH Lee, BG Moon and BS You, Tension–compression yield asymmetry in as-cast magnesium alloy. *Journal of Alloys and Compounds*, 2014. **617**: 277–280.
37. Shoji, H, The Bauschinger effect. *Zeitschrift fur Physik*, 1928. **51**(9–10): 728–729.
38. Sowerby, R, DK Uko and Y Tomita, Review of certain aspects of the Bauschinger effect in metals. *Materials Science and Engineering*, 1979. **41**(1): 43–58.
39. Pedersen, OB, LM Brown and WM Stobbs, The Bauschinger effect in copper. *Acta Metallurgica*, 1981. **29**(11): 1843–1850.
40. Bate, PS and DV Wilson, Analysis of the Bauschinger effect. *Acta Metallurgica*, 1986. **34**(6): 1097–1105.
41. Levine, LE, MR Stoudt, A Creuziger, TQ Phan, RQ Xu and ME Kassner, Basis for the Bauschinger effect in copper single crystals: changes in the long-range internal stress with reverse deformation. *Journal of Materials Science*, 2019. **54**(8): 6579–6585.
42. Abel, A and H Muir, Bauschinger effect and discontinuous yielding. *Philosophical Magazine*, 1972. **26**(2): 489–504.
43. Male, AT and MG Cockcroft, Method for determination of coefficient of friction of metals under conditions of bulk plastic deformation. *Journal of the Institute of Metals*, 1964. **93**(2): 38–46.
44. Sofuoglu, H and J Rasty, On the measurement of friction coefficient utilizing the ring compression test. *Tribology International*, 1999. **32**(6): 327–335.
45. Robinson, T, H Ou and CG Armstrong, Study on ring compression test using physical modelling and Fe simulation. *Journal of Materials Processing Technology*, 2004. **153**: 54–59.
46. Zhu, YC, WD Zeng, X Ma, QG Tai, ZH Li and XG Li, Determination of the friction factor of Ti-6Al-4V titanium alloy in hot forging by means of ring-compression test using FEM. *Tribology International*, 2011. **44**(12): 2074–2080.
47. Cristino, VAM, PAR Rosa and PAF Martins, The role of interfaces in the evaluation of friction by ring compression testing. *Experimental Techniques*, 2015. **39**(4): 47–56.

7 Hardness Testing

Hardness test procedures of various types have been in use for many decades. They are usually quick and easy to carry out, the equipment required is relatively simple and cheap, and there are portable machines that allow in situ *measurements to be made on components in service. The volume being tested is relatively small, so it's possible to map the hardness number across surfaces, exploring local variations, and to obtain values from thin surface layers and coatings. The main problem with hardness is that it's not a well-defined property. The value obtained during testing of a given sample is different for different types of test, and also for the same test with different conditions. Identical hardness numbers can be obtained from materials exhibiting a wide range of yielding and work hardening characteristics. The reasons for this are well established. There have been many attempts to extract meaningful plasticity parameters, particularly the yield stress, from hardness numbers, but these are mostly based on neglect of work hardening. In practice, materials that exhibit no work hardening at all are rare and indeed quantification of the work hardening behavior of a metal is a central objective of plasticity testing. The status of hardness testing is thus one of being a technique that is convenient and widely used, but the results obtained from it should be regarded as no better than semi-quantitative. There are procedures and protocols in which they are accorded a higher significance than this, but this is an unsound approach.*

7.1 Concept of a Hardness Number (Obtained by Indentation)

Reviews are available [1, 2] that summarize the historical development of hardness testing. Systematic attempts to characterize the hardness (resistance to plastic deformation) of materials can be traced back [2] to the proposal in 1812 by the Austrian mineralogist Friedrich Mohs that the capacity of one material to scratch another could be used as a basis for a ranking order (see §7.4.2). Suggestions of using a single hard indenter on a range of metals date back [2] to the work of William Wade in 1856, oriented towards optimizing materials for production of cannons. Commercial set-ups for testing hardness in this way started to become available around the beginning of the twentieth century. It was, however, several decades before serious attempts were made to establish a sound theoretical background to this type of testing [3–5], with the work of David Tabor being particularly notable [6, 7]. A small number of handbooks on hardness testing are also available, such as the compilation of chapters on individual tests edited by Herrmann [8].

Hardness is a measure of the resistance that a material offers to plastic deformation. It's of interest to have information, not only about the yield stress, but also about the subsequent work hardening characteristics. The hardness number provides a yardstick that incorporates both, although not in a well-defined manner. In view of the complexity of what it represents, it's unsurprising that hardness is not a simple, well-defined parameter and there are several different hardness measurement schemes, each giving different numbers. The idea, however, is the same for all of these schemes. A specified load is applied to an indenter, which penetrates into the specimen, causing plastic deformation and leaving a permanent depression. A hardness number can be obtained in several ways, but in most cases this is either via measurement of the indent lateral size (diameter) or of the penetration depth.

In principle, the time for which the load is applied should not affect the outcome. Practice is often rather different, since creep deformation (§4.5) can sometimes occur during ramping up and maintaining the load. Certain test procedures therefore include specification of a dwell time, for which the peak load is maintained. This is a concern, because plasticity characteristics (stress–strain curves) are in general regarded as independent of time, but such procedures do at least represent an attempt to improve the reproducibility. Leaving aside any creep effects, the depth and shape of the depression depend on the load, the shape of the indenter and the response (hardness) of the specimen. The indenter itself, which is commonly made of diamond, a cermet or a hard steel, should not undergo any plastic deformation.

Hardness is in most cases defined as the force (load) divided by the area of contact between indenter and specimen (although this is not the case for all schemes – for example, see §7.2.2). This ratio has dimensions of stress, although it is usually quoted as simply a number (with units of kgf mm^{-2}). In any event, this stress level bears no simple relation to the stress–strain curve, or indeed to the stress field created in the sample. Different regions of the specimen will have been subjected to different plastic strain levels, ranging from zero (at the edge of the plastic zone) to perhaps several tens of % (close to the indenter). Even this maximum strain level is not well defined, since it depends on the indenter shape, the applied load and the plasticity characteristics. While the stress–strain relationship of the material (von Mises stress as a function of equivalent plastic strain – see §3.1.1 and §3.1.2) does dictate the indent dimensions (for a given indenter shape and load), inferring the former from the latter is not straightforward and no attempt is made to do this in conventional hardness testing. (Chapter 8 concerns indentation plastometry, which does allow this to be done, although indents made using conventional hardness machines are not well suited to this methodology.)

Indenter shapes can be grouped into two broad classes, on the basis of whether or not they are "*self-similar.*" A self-similar shape is one for which the geometry of an axial section through the indenter (and sample) remains unchanged, apart from its scale, as penetration occurs. Characteristics of such shapes include the fact that the ratio of contact area to depth is constant (for a given shape, and ignoring effects of "*pile-up*" or "*sink-in*" around the indenter). There is thus no "scale effect." This is sometimes considered advantageous for certain types of analysis, mainly those based on load–displacement data. Most indenter shapes in which the sides are linear in an axial section are self-similar. The most commonly encountered indenter shape that is not self-similar is that of a sphere, for which the ratio of contact area to depth is given by $2\pi R$, where R is the radius of the indenter.

A related concept is that of the "*area function*" of an indenter [9]. This gives the actual contact area, as a function of the depth, for a particular shape, potentially taking account of factors such as elastic recovery and the effect of pile-up or sink-in, and perhaps also the actual shape right at the tip of a "sharp" indenter. It can be estimated in various ways for shapes that are self-similar (in which case the area/depth ratio should be constant on a simple geometric basis) and for those that are not. The attraction of this is that, with depth-sensing equipment, the contact area (and hence the hardness) can be obtained without the need for measurement of lateral dimensions of the indent, which is the "traditional" approach to measurement of hardness. In practice, both self-similar and non-self-similar indenter shapes are used in hardness testing. A brief description is provided below of the main tests (in chronological order of their development).

7.2 Indentation Hardness Tests

7.2.1 The Brinell Test

Johan August Brinell proposed a hardness test in 1900, while working at the Fagaresta Ironworks, before he became Chief Engineer of the Swedish Ironmasters Association. His test has remained in use ever since. It involves pushing a 10 mm diameter hardened steel or tungsten carbide (cermet) sphere into the sample, using a 3000 kg (~30 kN) load. It is oriented towards steels and is not suitable for much softer metals. A photo of a typical Brinell testing machine is shown in Fig. 7.1. At the front in the upper part is a device for measuring the indent diameter (via an optical microscope). The machine shown is a relatively old one – newer versions often incorporate more advanced imaging capabilities.

The Brinell hardness number is given by

$$H_B = \frac{2F}{\pi D \left[D - \sqrt{(D^2 - d^2)} \right]} \tag{7.1}$$

where F is the applied load (in kgf), D (mm) is the diameter of the indenter and d (mm) is the diameter (in projection view) of the indent. This formula corresponds to the load divided by the contact area (with the units being kgf mm^{-2}) and indeed this is how most hardness numbers are defined. It may be noted at this point that such formulae are based on a simple geometric approach. Elastic recovery of the specimen is neglected. Furthermore, in practice there may be "pile-up" or "sink-in" around the

Fig. 7.1 Photo of a typical Brinell hardness testing facility.

indent, such that the true area of contact differs from that obtained from idealized geometry (and also making accurate measurement of the diameter difficult – see below). No account is normally taken of such effects.

It should also be appreciated that typical numbers obtained in different types of (indentation) hardness tests are significantly different. This is unsurprising in view of the dependence of the plastic strain field on the indenter shape and the applied load, as well as on the plasticity characteristics of the material. One of the issues with hardness testing is that, since the load does tend to affect the hardness number, it should always be provided when quoting one (but often is not). At least with the Brinell test this is unnecessary (since the load is normally fixed). It is a high load, so a substantial loading frame and load generation system (often hydraulic) are required.

One of the attractions of the Brinell test in the early days was the associated claim that dividing the hardness number by 2 gave the UTS in units of ksi (thousands of pounds force per square inch). The conversion factor from ksi to MPa is about 6.9, so H_B would need to be multiplied by about 3.5 to obtain this "strength" in MPa. Even with recognition that this was only ever intended to be approximate, and that it might be a useful guideline if restricted to a fairly narrow range of steels, in reality no such conversion is likely to be at all reliable. It is not even expected to be accurate in terms of a ranking list.

One rather ironic aspect of the Brinell test, given that it was more or less the first one, is that, while the limitations relating to obtaining well-defined plasticity characteristics from it do apply equally to all hardness tests, it does at least involve the mechanical interrogation of a relatively large (and hence "representative") volume of the sample. The large (10 mm) diameter of the ball ensures that, at least in the vast majority of cases, this volume will contain many grains. It was recognized at an early stage that this was an advantage for testing of "heterogeneous" materials. However, the large indent size was also regarded as a disadvantage, because it meant that the test couldn't really be described as "non-destructive." When applied to a component in service, the residual indent was (and is) seen as potentially a significant defect. It also requires a set-up with a relatively high load capability. Most of the other hardness tests, and indeed much of the more recent indentation work generally, have moved to creation of progressively smaller indents. This has often had the effect of creating outcomes that cannot be regarded as representative of the plasticity of the bulk material – for example, the indents may lie within single grains, or perhaps just deform a small assembly of them. It's worth noting in this context that the indentation plastometry technique (Chapter 8) involves use of an indenter with a diameter (~1–2 mm) that is large enough to ensure that the outcome reflects the bulk plasticity, at least in most cases, but creates much smaller indents than the Brinell test, and hence can safely be regarded as "non-destructive."

A number of fairly complex mathematical treatments [10–12] have been published relating to the Brinell test (and indeed to some of the other tests as well), covering both analytical treatments and the outcomes of FEM modeling. Unfortunately, obtaining analytical solutions of any sort concerning such tests usually requires rather unrealistic assumptions, such as neglect of elastic recovery, work hardening, interfacial friction, surface topographies etc. Sensitivities are often such that qualitatively incorrect deductions can be made when such simplifications are employed. FEM modeling, in which it's not necessary to impose unrealistic boundary conditions, and for which the (true) stress–strain relationship employed can be expressed either as a set of data-pairs or as a constitutive law, does not suffer from these limitations. It therefore tends to be the most effective way of investigating the actual behavior of a sample during such a test, although the identification of underlying characteristics can then be a rather cumbersome operation.

Some outcomes of such modeling of the Brinell test are now shown, based on the two stress–strain curves shown in Fig. 7.2 (approximately corresponding to the two types of steel indicated). One point of interest is what hardness numbers would be obtained from these indents, if measured in the conventional way. Profiles are shown in Fig. 7.3 for these two steels, with three different applied loads. As for most indents, there is a degree of pile-up (greater for metals exhibiting less work hardening – i.e. for the duplex stainless in this case). In both cases, however, the formation of a pile-up (or sink-in) raises the question of what value would in practice be obtained for the indent diameter, d, given that there is not a well-defined "rim" to the "crater." In Fig. 7.3, an attempt has been made to identify the locations that would probably be perceived as the "edges" of the indents. If the values shown are used in Eqn. (7.1), then, for the Mangalloy, the H_B numbers obtained would be 189, 205 and 210, for loads of 10, 20 and 30 kN, respectively. For the duplex stainless, corresponding H_B values are 97, 106 and 105. In fact, there is always likely to be uncertainty about the measured diameter, with sensitivities such that the associated error in hardness number could be

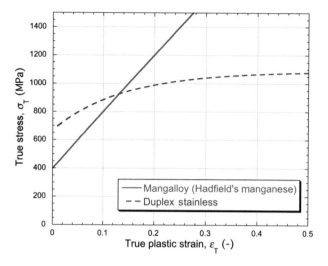

Fig. 7.2 True stress–strain curves, corresponding to specific constitutive laws, intended to be representative of two different types of steel.

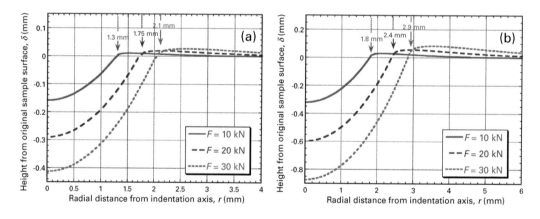

Fig. 7.3 FEM-predicted indent profiles, after application of the forces shown, using a (Brinell) spherical indenter of diameter 10 mm, for materials with the true stress–strain relationships in Fig. 7.2, corresponding to (a) Mangalloy and (b) duplex stainless steel. An indication is given of values likely to be obtained for the radius (= $d/2$) of the indents, via optical microscopy.

relatively large. For example, if the measured radius for the largest Mangalloy indent had been recorded as 2.0 mm, rather than 2.1 mm, then H_B would be raised from 210 to 230.

It may also be noted that there is an explanation for the observed rises in H_B with increasing load. For example, Fig. 7.4 shows fields of von Mises strain for the Mangalloy, after application of the three loads. It can be seen that, as expected, the peak strains, and other strains close to the indenter, rise with increasing load, although all of these strains are relatively low (since this is quite a hard steel). A useful concept

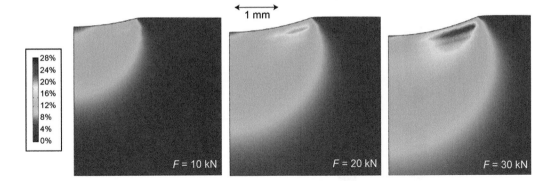

Fig. 7.4 FEM-predicted fields of equivalent plastic strain in a Mangalloy sample, with the true stress–strain relationship shown in Fig. 7.2, after Brinell testing with the applied loads shown.

when aiming to understand the effect of these strain fields on the mechanical response is that of the average plastic strain during indentation, weighted for the amount of plastic work done in different ranges of strain [13]. More information about the procedure for evaluating these strains is provided in §8.4.1. For these three loads, the average plastic strains are, respectively, 3.9%, 5.3% and 6.4%. These are relatively low levels, both because the penetration is not very deep and due to the high work hardening rate. The results shown in Fig. 8.4 confirm that values between 10% and 15% are common, for a typical penetration ratio (δ/R) of 20%.

Nevertheless, since the Mangalloy does exhibit quite strong work hardening (Fig. 7.2), it is expected that higher average plastic strain values will raise the hardness (resistance to indenter penetration). Of course, since Brinell testing is normally carried out with a single load (3000 kgf), this will not in practice be a source of variation, but the point still holds for other tests in which a range of loads is routinely used, such as the Vickers test (§7.2.3).

7.2.2 The Rockwell Test

There are many issues with hardness set-ups, including a potential influence of the surface roughness of the sample and the possibility of the mechanical "backlash" of the machine affecting the outcome. These were recognized at an early stage and led to the idea of a multi-stage test – i.e. to a preloading of some sort, before the test proper was carried out. A patent covering this idea was filed in 1914 by the Rockwell brothers (Hugh and Stanley). Stanley Rockwell subsequently founded various metallurgical and testing firms, part of which was acquired by the Instron Corporation in 1993.

Two types of indenter are used in Rockwell testing. One is a diamond cone with a 120° included angle. There is always an issue with such geometries concerning the exact shape at the tip. This is taken to be spherical (and indeed there must always be a finite radius at the tip of any such indenter), but the radius of the tip region is not

Fig. 7.5 Photo of a typical Rockwell hardness testing facility.

defined. The other type of indenter used is a (hard) steel sphere of diameter 1.588 mm (1/16 inch). This at least has a well-defined geometry. The diamond cone is used for hard materials, such as tungsten carbide (Rockwell scale A) and relatively hard steels (Rockwell scale C). The steel sphere is used for softer materials, such as Al, Cu, brass and less hard steels (Rockwell scale B).

A photo of a typical Rockwell hardness tester is shown in Fig. 7.5. The procedure involves applying a pre-load ("minor load"), which is normally 10 kgf (98.1 N). The depth of the resulting indent will, of course, vary with the hardness of the material, but it is usually at least several tens of microns, so that it penetrates the region of surface roughness, oxide films etc., at least in most cases. The penetration measurement device, which is usually some kind of dial gauge, is set to zero. The main load is then applied. This is an additional 140 kgf for Rockwell C, 90 kgf for Rockwell B and 50 kgf for Rockwell A. (This latter value is smaller, despite tungsten carbide being in the hardest category, mainly to reduce the danger of damaging the diamond tip.) This main load is usually applied for a specified time, or possibly until the dial gauge "stops moving." The main load is then removed. The penetration distance, δ, used to obtain the hardness number, is the difference between this depth (with the minor load still applied) and the depth when the minor load was applied originally (at which the dial gauge had been zeroed).

The hardness number for category C is then given by

$$H_{\text{RC}} = 500(0.2 - \delta) \tag{7.2}$$

where δ is in mm. This looks, and indeed is, rather arbitrary. It doesn't really represent an attempt to estimate the load over the contact area, which is handicapped by uncertainty about the exact geometry of the tip of the diamond indenter. In practice, it's just an empirical correlation. Of course, it could be argued that, since any hardness number has little or no intrinsic meaning, any number will do: it will at least be larger for cases in which the indenter penetration is lower (and the material is harder). There is no pretense of relating it to a yield stress or any work hardening characteristics. It can be seen that an "infinitely hard" material (no penetration) will give a value of 100. In practice, the Rockwell C hardness values for a range of (fairly hard) steels run from about 10 up to around 70.

For category B, the equation used is

$$H_{RB} = 500(0.26 - \delta) \tag{7.3}$$

Again, this is fairly arbitrary. In particular, it does not arise from a geometrical construction aimed at equating the hardness number to the load over the contact area (as the Brinell number does). The two-stage loading procedure means that the contact area cannot be simply expressed in terms of δ, even when the sphere diameter is known (as it is for the Rockwell B category).

The main attraction of Rockwell testing is that there is no need to measure (optically) an indent diameter. The depth is measured by the machine and automatically converted to a hardness value (usually displayed directly on the dial gauge). Also, the idea of a two-stage loading procedure does have theoretical advantages in terms of eliminating the effects of surface roughness etc. It is common to see charts allowing conversion between hardness numbers for the different schemes, but these also are simply empirical correlations. It's clear from the rather arbitrary nature of all hardness numbers that such conversions give only guidelines at best.

7.2.3 The Vickers Test and Berkovich Indenters

The Vickers test was originally developed in 1924, by Smith and Sandland (at Vickers Ltd.). A key objective in devising this test was to reduce the load requirements. Both the Brinell (3000 kg) and the Rockwell (~100 kg) tests require loads that are too high to be readily supplied via simple dead-weights located inside the machine. Hydraulic loading systems are commonly employed for them. Changing the indenter from a relatively large sphere (or cono-spherical shape) to a smaller and "sharper" shape allowed a lower load (that could be created with a dead-weight) to be used. Several such weights are usually provided inside the machine, ranging from below 1 kg up to around 50 kg, depending on the model. The (diamond) indenter is a right pyramid with a square base and an angle of 136° between opposite faces. The (sharp) edges promote penetration and the lines that they produce in the indent facilitate measurement of its size.

A photo of a typical Vickers testing machine is shown in Fig. 7.6(a) and the geometry of the indenter and indent is illustrated in Fig. 7.6(b). The measured indent

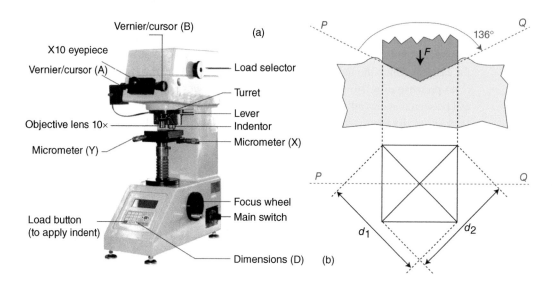

Fig. 7.6 The Vickers hardness test: (a) photograph of the equipment and (b) geometry of the indent.

diameter, d, taken as the average of d_1 and d_2, is measured in projection (as for the Brinell test). The value of H_V (load divided by contact area) is given by

$$H_V = \frac{2F \sin\left(\frac{136}{2}\right)}{d^2} = 1.854 \frac{F}{d^2} \tag{7.4}$$

A simple calculation, similar to that for the Brinell test (Eqn. (7.1)), thus allows the hardness number to be obtained from the measured value of d. As with the Brinell test, elastic recovery of the specimen, and "pile-up" or "sink-in" around the indent, are neglected.

The Vickers test is quite widely used. In fact, H_V is probably the most commonly quoted of the hardness numbers, partly because, by varying the load, it can be applied to a wide range of metals, and to thin sections, surface layers etc. Also, the equipment required is cheap and simple. Figure 7.7 shows a typical set of values [14], covering various alloys. These were obtained via a careful set of measurements on indent dimensions in particular samples, although in practice these values would change in many cases if the thermomechanical treatments had been modified. These data do serve to illustrate typical ranges, although the exact numerical values should, to say the least, be treated cautiously.

The stress acting on the contact area (in MPa) is obtained on multiplying this hardness number by g (9.81). This stress does not bear any simple relation to the stress–strain curve. However, if work hardening is neglected, then the hardness should be proportional to the yield stress (i.e. the "flow stress"). For the Vickers test, the relationship is often written as

$$\sigma_Y = \frac{H_V}{3} \tag{7.5}$$

7.2 Indentation Hardness Tests

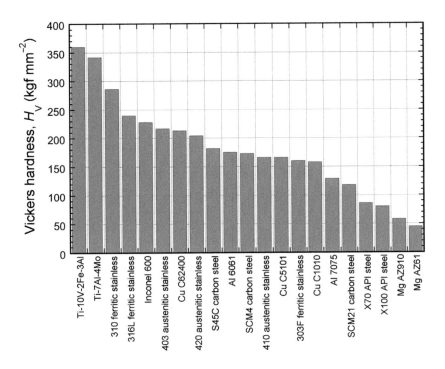

Fig. 7.7 Data [14] for the Vickers hardness of a range of alloys.

Such expressions are commonly used to obtain a yield stress from a hardness measurement and the background for this kind of estimate should be understood. It is based on the broad nature of the stress and strain fields expected under a Vickers indenter, in the absence of work hardening. This is illustrated here by some FEM outcomes, for a material with a yield stress of 250 MPa (and no work hardening), using a "Vickers equivalent" cone, which has a semi-angle of 70.3°. (This is done, rather than carrying out a full 3-D analysis with an actual Vickers pyramid, simply to make it easier to visualize the outcomes.) Figure 7.8 shows predicted residual indent profiles for three applied loads, with indications of the indent radii that would be obtained experimentally – these are the measurements that would correspond to the d value in Eqn. (7.4). The contact area in this case, based on the same kind of simple geometrical approach, is given by

$$A = \pi r \left(r^2 + h^2\right)^{1/2} = \pi r \left(r^2 + \frac{r^2}{\tan^2(70.3°)}\right)^{1/2} = 3.337\, r^2 \qquad (7.6)$$

where h is the depth of penetration. The equivalent expression for the hardness (force over contact area) in this case is thus

$$H_{VC} = \frac{F}{3.337 r^2} \qquad (7.7)$$

with H_{VC} having units of MPa if F is in N and r is in mm.

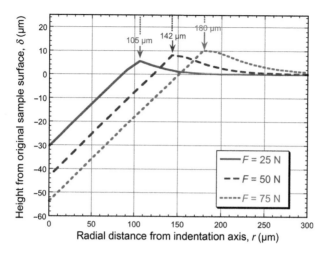

Fig. 7.8 Predicted (FEM) residual indent profiles after indentation with a "Vickers equivalent" cone, to three different loads, of a material with a yield stress of 250 MPa and no work hardening.

Using the r values in Fig. 7.8 (taken to be those that would be measured experimentally), and dividing H_{VC} by 3 as in Eqn. (7.5), gives σ_Y values of 226 MPa, 248 MPa and 231 MPa, respectively, for the loads of 25 N, 50 N and 75 N. These figures should be compared with 250 MPa for the material being modeled. This is all very approximate, even for a case in which there is no work hardening, and the actual indents are some way from simply reflecting the geometry of the cone. For example, the actual depth below the far-field free surface for the 75 N case is about 54 µm, whereas, geometrically, for the measured r value of 180 µm, it should be about 64 µm. This difference arises from elastic recovery and the effect of the pile-up. In any event, these results provide an indication of how approximations such as Eqn. (7.5) have arisen. Further insights can be obtained from Fig. 7.9, which shows the (von Mises) stress and plastic strain fields under the indenter for the 75 N applied load. It can be seen that there is a large region in which this stress is equal to the yield stress. Also, the plastic strains have reached high levels (~170%) close to the apex of the cone.

Furthermore, the caveats mentioned above should be reiterated here. The assumption of "perfect plasticity" is not normally acceptable and even relatively minor work hardening will change this behavior considerably. If a yield stress value obtained in this way is regarded as some sort of average "flow stress" over the range of plastic strain created during the test, then the error may not be so great, but it is not a simple matter to estimate this range [15, 16] (and it will be different for different applied loads). There have been various attempts [17, 18] to treat the Vickers test analytically, but the difficulties are even more daunting than with spherical indenters. Of course, (3-D) numerical (FEM) simulations can be carried out [19, 20].

There is also another concern about this test, which does not apply to the same extent for Brinell and Rockwell tests. Taking the example of a 10 kg load and a sample

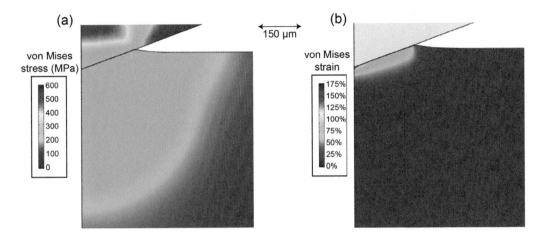

Fig. 7.9 Predicted (FEM) fields of (a) von Mises stress and (b) von Mises strain during indentation with a "Vickers equivalent" cone at an applied load of 75 N, of a material with a yield stress of 250 MPa and no work hardening.

having an H_V value of 200 (a typical medium strength steel or a Ti alloy), use of Eqn. (7.4) indicates that the diameter of the indent is about 0.15 mm (150 μm). The corresponding depth is about 20 μm. This size range is rather similar to that of many grain structures and the possibility starts to arise of the indent being located within a single grain, or perhaps a small group of grains. As emphasized elsewhere (§4.3.1 and §5.1.3), if bulk properties are required, then the assembly of grains being mechanically deformed must be relatively large. Of course, a larger load can be used for harder materials, in an attempt to ensure that the deformed volume is sufficiently large to be "representative," although with many Vickers machines the maximum load is 10 kg. This effect is often manifested during testing as a large degree of scatter between repeat tests in different locations, although it is important to appreciate that the "correct" mechanical response is not simply the average of those from a number of (differently oriented) individual grains.

In fact, it has become common to carry out Vickers indentation with quite low loads, often as part of an activity that is usually described as "nanoindentation" – see Chapter 9 for details. The key characteristic of such indentation is not really that it is carried out on a very fine scale, although of course it often is, but rather that there is continuous (and very accurate) monitoring of the displacement (and load) during the test. While obtaining a hardness value is often not the prime objective of "nanoindentation," it is frequently done – commonly via analysis of the load–displacement plot [21], rather than from measurement of the lateral dimensions of the indent. However, there are certainly dangers in obtaining hardness values from fine scale indentation. One of them centers on the issue of a "representative volume" outlined above. Another relates to the fact that, as described at the end of §8.2.2, plastic strain levels will vary (being lower for smaller indents), changing the parts of the stress–strain curve to which the outcome is sensitive. (The concept of a "representative

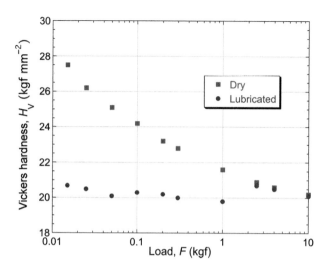

Fig. 7.10 Effects of load and lubrication on the measured Vickers hardness of annealed aluminum [22].

strain" in the Vickers test has been addressed in the literature [15, 16].) A third issue concerns the increasing significance of surface roughness and oxide film effects as the penetration depth is reduced. Of course, it can readily be argued that a hardness number is in any event of limited inherent significance, but there is certainly a likelihood of a strong "size effect" emerging as indents become very small, such that it becomes even more meaningless.

This is illustrated by the data in the plot [22] of Fig. 7.10, which shows values of H_V obtained (for the same material) over a range of applied loads, with and without lubrication. Various explanations have been put forward for such effects, which often (but not always) take the form of increasing hardness as the indentation depth is reduced. It should in any event be recognized that data of this type often exhibit a lot of scatter when measurements are repeated. The effect of lubrication apparent in Fig. 7.10 suggests that surface-related phenomena are playing an important role for fine indents – of course, lubrication is uncommon in normal practice. The material concerned is very soft, but nevertheless the penetration created by loads of 10–100 g (0.1–1 N) is only of the order of a few microns, although it may be noted that much "nanoindentation" is carried out to depths that are substantially sub-micron: the effects of surface roughness, contamination and oxide films are expected to be very strong in this regime.

Finally, a point can be made that applies to virtually all "sharp" indenters, which relates to the danger of them becoming damaged in some way. "Edges" and "points" are always prone to such damage, since the stresses created at them, and also the plastic strains in the material in contact with them, tend to be relatively high. These also tend to be regions where oxidative attack is more concentrated, which is a factor to bear in mind with diamond indenters (if used at high temperature). Of course, any such damage is likely to affect the hardness readings obtained using that indenter.

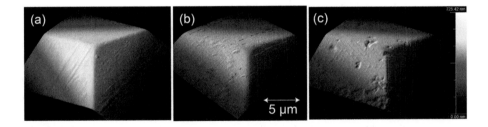

Fig. 7.11 AFM reconstructions [23] showing Berkovich tips that had been exposed to: (a) extensive usage (below 400 °C), (b) 30 minutes at 450 °C in air and (c) 90 minutes at 900 °C in Ar (~40 ppm of O_2).

As an illustration of the type of damage that can occur, three AFM (atomic force microscope) topographic images [23] are shown in Fig. 7.11. These are of Berkovich tips exposed to different service conditions. Berkovich tips are similar to Vickers tips, but based on a three-sided pyramid, rather than a four-sided one. (The main motivation for using a Berkovich, rather than a Vickers, is actually that it is much easier to produce: grinding a diamond to create a single sharp point is easier for a three-sided pyramid than for a four-sided one.) They are popular for use in depth-sensing "nanoindenters" (see §9.2.2). It can be seen from these images that diamond tips of this type can become damaged, particularly if exposed to high temperature in the presence of oxygen. The oxidation rate is sensitive to both temperature and oxygen partial pressure [23], although loss of sharpness at the edges can be seen even with the tip (Fig. 7.11(a)) that had not been heated substantially, but had been in use for an extended period. Spherical indenters are much less susceptible to damage during use.

7.2.4 The Knoop Test

The most recent of the "standard" set of hardness tests is that of Knoop, which was proposed [24] by Frederick Knoop in 1939. The shape of the indenter is similar to that of Vickers – i.e. it is a four-sided pyramid – but the length of the long diagonal is about seven times that of the short diagonal. This unusual shape is designed to create relatively shallow penetration (while still having a lateral dimension that is relatively large and hence easy to measure). Arguments have also been put forward [24, 25] in terms of this indenter shape creating less pile-up than other indenters, so that it is easier to establish the actual area of contact and hence to obtain a more "accurate" hardness number. An obvious counter-argument is that, since there is no inherent significance in evaluating the contact stress, and calling it a hardness number, this can't really be regarded as an improvement in accuracy: even if a contact pressure is accurately established, hardness numbers defined in this way are very different for different indenter shapes and loads.

The Knoop hardness number is given by

$$H_K = \frac{F}{CL^2} \tag{7.8}$$

where L is the measured length (in projection) of the long diagonal and C is a factor related to the indenter shape. In principle – i.e. from the actual geometry of the indenter – it has a value of about 0.07. Values of H_K are typically in the range 100–1000. It is often used for relatively hard materials (including ceramics) and for thin layers, surface coatings etc. Loads applied during Knoop testing are often relatively low (<1 N) and indents are very shallow. This immediately raises the issue of surface roughness/contamination effects and surfaces often need careful penetration. Also, even the long diameter is commonly very short and difficult to measure accurately (via optical microscopy). Furthermore, very low loads of this type are difficult to apply using the "traditional" methods of hardness testing (dead-weights and hydraulic systems).

Development of the Knoop indenter is actually part of a general movement towards finer scale indentation, and also towards "***depth-sensing indentation***," although the latter type of capability (commonly described as "nanoindentation") was not really available at the time that the Knoop test was proposed. In fact, Knoop indenters, and also Vickers and Berkovich indenters, are now most commonly used as part of such a facility. Moreover, (Knoop) hardness numbers are often obtained, not via measurement of an indent size, but rather by analysis of the load–displacement plot recorded during the test. (A similar type of analysis is also commonly used to obtain the Young's modulus of the material, and indeed this is a well-defined and fairly accurate procedure – see §9.3.2: unfortunately, since hardness is not a well-defined property, it cannot be obtained "rigorously" by any procedure.) This is often done automatically in some way, with associated risks in terms of the danger of such a procedure masking exactly how the penetration is occurring. Such issues – i.e. the procedures and analyses used during "nanoindentation" – are covered in Chapter 9.

7.3 Effects of Sample Condition

7.3.1 Microstructure, Anisotropy and Indentation of Single Crystals

The trend towards finer scale indentation, while experimentally attractive in many ways, raises issues related to the scale of the microstructure. This particularly concerns the grain structure, as described in §7.2.3, but it also relates to smaller features such as precipitates etc. Indeed, an attractive feature of "nanoindentation" is that it allows the hardness of such phases to be established. This would otherwise be difficult in many cases (since it is often a major challenge to create large samples of such phases). Of course, this doesn't change the fact that hardness is an ill-defined property, but it is nevertheless a useful capability to be able to monitor the mechanical characteristics (including the stiffness) of very fine constituents in this way. There may, however, still be issues of the surface roughness, and contamination/oxide layers, which can affect the outcome.

There are certainly complicating factors when creating indents within individual grains, or when indenting a single crystal. The latter is in some ways an easier

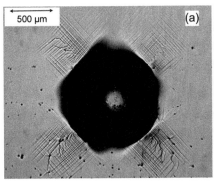

Fig. 7.12 Optical micrographs of the free surfaces of samples that had been indented using a 1 mm radius cermet sphere, for (a) a Ni superalloy single crystal and (b) a copper sample, with a grain size of about 100 μm. (Some cracking can be seen in the superalloy sample.)

situation to treat than the former, since it will in general be easier to establish the orientation of a single crystal than an individual grain within a polycrystal. In addition, with a single crystal there need be no concern about the possibility that the indent could be straddling a grain boundary, or deforming a small number of grains – a scenario that would be very difficult to interpret in a systematic way.

Figure 7.12 shows optical micrographs of the free surfaces of samples after indentation with a spherical indenter of radius 1 mm. The sample of Fig. 7.12(a) is a single crystal, while Fig. 7.12(b) is from a typical polycrystalline specimen (at higher magnification). A number of important differences are clearly apparent, although it may be noted that both show a large number of "surface steps." These are ***persistent slip bands*** – intersections of the free surface with slip planes along which large numbers of dislocations have glided. Information about dislocation motion is provided in §4.2 and §4.3. For the single crystal, which was oriented with a <100> direction normal to the free surface, the slip occurred on a well-defined set of planes (of (111) type). It can be seen that one consequence of this is that the indent does not exhibit radial symmetry. Put another way, the material is plastically (and elastically) anisotropic. It would be possible to analyze this situation, and perhaps obtain information about this particular crystal (such as the critical shear stress for dislocation glide – the Peierls stress), but it would need to be done with a knowledge of the crystal orientation and with account taken of which slip systems were operational. A single hardness value could not be obtained (since it varies with direction in the crystal).

With the indent of Fig. 7.12(b), on the other hand, it is relatively easy to obtain information about (bulk) plasticity characteristics, despite the fact that what actually happens during indentation is considerably more complex than for the single crystal. A large assembly of grains was deformed during the indentation, so the response is representative of the bulk (the behavior of which depends on grain size, texture, grain boundary structure etc., as well as on intra-granular features such as purity, alloy composition, precipitate size and dispersion etc.). It can be seen that multi-system slip occurred in most of the grains, as it does during tensile testing (§4.3.1). There is also evidence of (small) rotations of individual grains – it may be noted that the plastic strains in all of these regions at the free surface are relatively low (<~1–2%). Furthermore, at least in most cases, the response is isotropic, so the indent exhibits radial symmetry and a similar outcome (indent profile) would be obtained if the material were to be indented in a different direction – indentation involves multi-axial interrogation of the mechanical response. In fact, it's possible, not only to obtain a single hardness number, but also to infer the stress–strain curve of the material (via indentation plastometry, as described in Chapter 8).

7.3.2 Residual Stresses

Residual stresses are common in many (metallic) samples. Some information about them is provided in §3.3. Since hardness testing involves examination of a near-surface region, and there can be no normal stress at a free surface, the initial (pre-test) stress state must be a biaxial one, and it will commonly be equal biaxial (transversely isotropic), in which case it is fully defined by a single value. The stress state could vary with depth, but in general any such variation will be small within the region being interrogated during an indentation test. A similar argument applies to any lateral variations.

It's clear that the presence of such in-plane stresses may affect the indentation response. This effect is likely to be noticeable if their magnitude is significant compared with the yield stress. There are two main issues to consider here. Firstly, the presence of such residual stresses will tend to affect the outcome of the test – for example, the hardness number obtained from it. This might be regarded as the introduction of an "error," although, in view of the status of a hardness number, and the likely magnitude of the effect, this is probably of little concern in most cases. The second point to note is that there may be potential here for using indentation to evaluate the residual stress(es). This is an important point, since knowledge of the (near-surface) residual stress state, and of its variation over the surface of a component, is often an important objective. Most other techniques for obtaining such information are cumbersome (and rather inaccurate), and their lateral resolution capabilities are relatively poor. In fact, as a variant of the indentation plastometry technique, there is real potential in this area, as described in §8.4.3. The treatment here is limited to a brief perspective on the effect of residual stresses during hardness testing.

It has long been clear that residual stresses have the potential to affect hardness test outcomes. For example, Fig. 7.13 shows data from 1952, indicating the effect of (uniaxial) residual stress (induced via the bending of a bar of a high carbon steel) on the measured Rockwell B hardness. The sense of the effect is as expected, with a tensile applied stress promoting yielding (as the indenter starts to apply a normal compressive stress), leading to a reduction in the apparent hardness. However, the reported effect is rather small, with even an imposed stress of 300 MPa inducing only a 5% change in hardness – well within the expected experimental error.

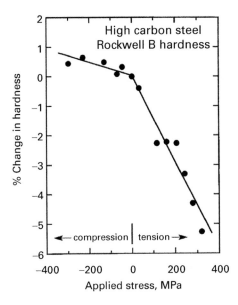

Fig. 7.13 Experimental data from a Sines and Carlson study of 1952, reported by Tsui et al. [26], concerning the effect of an applied stress on the measured Rockwell B hardness of a high carbon steel (obtained via indent dimensions).

There have since been a number of more systematic investigations [26–28] aimed at analysis of the effect of residual stresses on hardness test outcomes. Most of these have focussed on load–displacement plots (i.e. data obtained during depth-sensing indentation), rather than on lateral indent size measurements, which tend to be rather insensitive to these effects. For example, Tsui et al. [26] carried out experimental work using the set-up shown in Fig. 7.14(a). This created an unequal biaxial stress state – in fact, one that approximated to a uniaxial stress. They also created equal biaxial stress states, using a different experimental configuration.

A Berkovich indenter was used and the focus was on values obtained for the stiffness and the hardness (defined as the contact stress). These values were obtained from experimental load–displacement plots. The results are shown in Fig. 7.14(b) and (c). An Al alloy was used with a yield stress of about 350 MPa, and a very fine-grained microstructure (so that, even with lateral indent dimensions of about 20 μm, the mechanical interrogation was suitably multi-grained). It can be seen that significant effects were observed, with stress levels close to the yield stress apparently changing the hardness by about 10% and also inducing noticeable effects on the stiffness (although there is quite a lot of scatter in the data). Of course, the stiffness should not be affected by residual stress, so there are some issues about exactly how the load–displacement data should be handled in this situation. Tsui et al. [26], as well as later investigations focussed mainly on FEM simulation [27], have proposed explanations for the observed effects.

There have been other studies in this area, several of them [29–31] oriented towards use of Knoop indenters, with some claims that this shape offers advantages for this type of investigation. In fact, there are fundamental problems with any such "sharp"

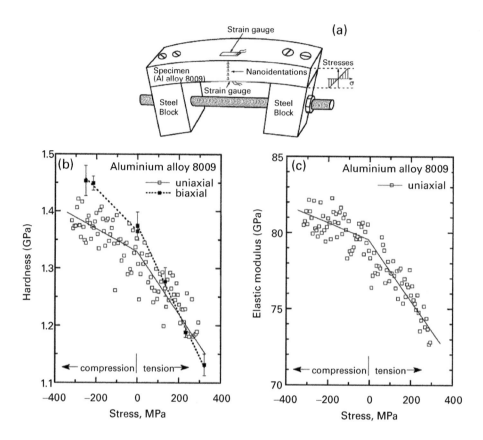

Fig. 7.14 Work of Tsui et al. [26] concerning the effect of imposed residual stresses on outcomes of hardness testing of an Al alloy with a Berkovich indenter, showing (a) the loading set-up for creation of a uniaxial stress, (b) measured hardness and (c) measured stiffness, both obtained from load–displacement plots.

indenters, since the stress and strain fields are not well defined close to the edges, so that iterative FEM modeling – which is the most promising approach to rigorous analysis – often suffers from poor sensitivity and accuracy. Furthermore, studies oriented towards hardness testing tend to rather miss the point when the most important objective is to use indentation testing to evaluate (unknown) residual stresses. (There have also been studies in which the grain size of the sample has been larger than typical indent dimensions, creating another source of variability.) In any event, there has been no "hardness testing" work in which a reliable methodology has emerged for doing this.

7.4 Other Types of Hardness Testing

7.4.1 Rebound Hardness Testing

Rebound hardness tests have attractions, particularly for *in situ* testing of components. The approach has its origins in the **Schmidt hammer test** [32] proposed in the late

1940s for testing of concrete, which has since been applied quite widely to rocks and minerals. The "hammer" is a spring-loaded piston and the way in which it rebounds after striking the sample is measured. However, the impact conditions are not very well defined in such a test and there has been a move, particularly for metals, towards using a free projectile, usually in the form of a hard sphere, still with the rebound behavior as the basis of the test.

An enclosure housing such a projectile is placed on the component. This ball, typically of hardened steel and with a diameter of about 2–4 mm, is then propelled onto the sample surface, usually with a velocity of a few tens of m s^{-1}. A device inside the enclosure measures the rebound velocity. With an "infinitely hard," perfectly elastic sample, incident and rebound velocities would be equal (and no permanent indent would be created). In practice, some plastic deformation occurs and there is a loss of (kinetic) energy equal to the plastic work done. As with all hardness testing, there is then the question of how the hardness number should be defined. This is necessarily arbitrary (with no relation to the idea of a contact stress used in defining most hardness numbers in indentation testing). Of course, rebound testing is an inherently dynamic process, often inducing relatively high strain rates – see below.

One of the procedures is that termed the **"Shore" test** [33], which is based on the ball falling under gravity (limiting the impact velocity). This restriction was removed by development [34] of the Leeb test (**"Leeb–Equotip durometer"**) in 1986. This is based on a spring-loaded projection system and is now widely used, particularly for metals. Both incident and rebound velocities are normally measured via electromagnetic induction.

The Leeb hardness number is defined as

$$H_L = \frac{1000 u_r}{u_i} \tag{7.9}$$

where u_i and u_r are the impact and post-rebound velocities. This ratio of post- and pre-impact velocities is often termed the **"coefficient of restitution."** It can be seen that an infinitely hard surface would lead to $H_L = 1000$, while a very soft material (no rebound) would give $H_L = 0$. Surface roughness is a particular issue with this type of test, since asperities tend to deform preferentially, with more plastic work being done than would be the case for a perfectly flat surface. The test specification normally requires a surface roughness below 10 µm, which can be demanding for a test carried out in the field. There is also an issue relating to the mass (and hence the inertia) of the sample, since energy can be absorbed in relatively low mass objects via the stimulation of vibrations (that become damped). Bodies with mass below about 5 kg are usually considered unsuitable.

Measured H_L values [35] are shown in Fig. 7.15, plotted against the yield stress of the material concerned (for a range of steels used for reinforcement purposes). Since these are all rather similar types of steel, the range of hardness covered is relatively small. It can be seen that, as expected, there is certainly some correlation with the yield stress, although there is quite a lot of scatter in the data. This is presumably due to variations in the work hardening characteristics. Also, while it might have been expected that the data would extrapolate to the origin (i.e. H_L would tend to zero

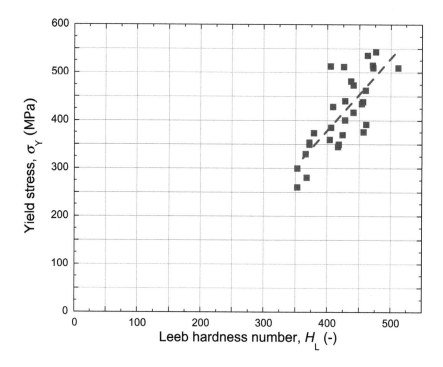

Fig. 7.15 Measured values [35] of the Leeb hardness, plotted against the yield stress, for various steels used to reinforce concrete.

for a very soft material), this does not appear to be the case (although there is certainly no reason why the relationship should be a linear one).

In fact, as with other hardness tests, the number obtained is not independent of the test conditions. It is often found to vary with the size and incident velocity of the ball. There are several possible reasons for such effects, but one that is likely to be significant is the **strain rate sensitivity** of the plasticity characteristics. The move towards higher incident velocities (and smaller projectiles) has led to the strain rates becoming relatively high. As outlined in §3.2.2 and §5.4.3, the stress–strain curve starts to deviate from its "quasi-static" version at high strain rates, typically becoming significantly different (i.e. a higher "flow stress" at a given strain) at rates above ~10^3 s^{-1}. Depending on projectile size, such rates are often created with impact velocities of the order of 10 m s^{-1}. In fact, tests of this type are sometimes used (in conjunction with FEM simulation) to study such strain rate effects, as described in §10.4.3. As far as simple rebound testing to obtain hardness values is concerned, however, this will be a source of error and variation.

7.4.2 Scratch Testing and the Mohs Scale

A number of procedures have emerged that involve a hard indenter being dragged across the surface of a sample. Furthermore, this is the basis of what is effectively the oldest of the approaches to hardness characterization – namely the idea put forward

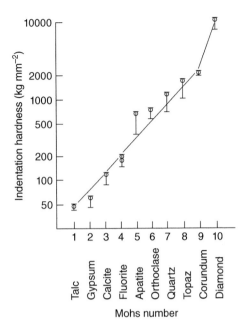

Fig. 7.16 Correlation [38] between the Mohs scale of hardness and Vickers hardness numbers.

early in the nineteenth century by Friedrich Mohs [36, 37] of a ranking order of materials based on their capacity to scratch each other. This has always been oriented mainly towards minerals, and indeed the Mohs scale is defined in terms of a set of 10 minerals. However, the range of hardness covered by the Mohs scale does extend down to values typical of many metals. This can be seen in the correlation [38] between the two shown in Fig. 7.16. Given that many metals have a (Vickers) hardness in the approximate range 100–300 kgf mm^{-2} – for example, see Fig. 7.7 – it can be seen that they fall within the Mohs scale range of about 1–4. Yet most of the Mohs range relates to harder materials and in general the scale is not applied to metals.

Nevertheless, **scratch testing** is quite widely used [39–41] on metallic samples. However, the processes that take place are inevitably complex, with a strong dependence on surface roughness, adsorbed layers, indenter geometry, motion speed, contact force and other test conditions. They are popular as empirical tests, partly because they are relevant to wear and abrasion characteristics, but it's difficult to relate the outcomes to fundamental (plasticity) parameters. Of course, harder metals are naturally more scratch-resistant than softer ones, but the sensitivity to test conditions is such that it's difficult to use a scratch test to infer a hardness, much less to obtain basic information about yield stress and work hardening characteristics.

References

1. Walley, SM, Historical origins of indentation hardness testing. *Materials Science and Technology*, 2012. **28**(9–10): 1028–1044.

2. Broitman, E, Indentation hardness measurements at macro-, micro-, and nanoscale: a critical overview. *Tribology Letters*, 2017. **65**(1).
3. Bishop, RF and NF Mott, The theory of indentation and hardness tests. *Proceedings of the Physical Society of London*, 1945. **57**(321): 147–159.
4. Tabor, D, *Hardness of Metals*. Oxford: Clarendon Press, 1951.
5. Mott, BW, *Microindentation Hardness Testing*. London: Butterworths, 1956.
6. Tabor, D, Indentation hardness: fifty years on – a personal view. *Philosophical Magazine A: Physics of Condensed Matter, Structure, Defects and Mechanical Properties*, 1996. **74**(5): 1207–1212.
7. Hutchings, IM, The contributions of David Tabor to the science of indentation hardness. *Journal of Materials Research*, 2009. **24**(3): 581–589.
8. Herrmann, K, *Hardness Testing: Principles and Applications*. Materials Park, OH: ASM International, 2011.
9. Herrmann, K, NM Jennett, W Wegener, J Meneve, K Hasche and R Seemann, Progress in determination of the area function of indenters used for nanoindentation. *Thin Solid Films*, 2000. **377**: 394–400.
10. Richmond, O, HL Morrison and ML Devenpeck, Sphere indentation with application to Brinell hardness test. *International Journal of Mechanical Sciences*, 1974. **16**(1): 75–82.
11. Hill, R, B Storakers and AB Zdunek, A theoretical study of the Brinell hardness test. *Proceedings of the Royal Society of London Series A: Mathematical, Physical and Engineering Sciences*, 1989. **423**(1865): 301–330.
12. Biwa, S and B Storakers, An analysis of fully plastic Brinell indentation. *Journal of the Mechanics and Physics of Solids*, 1995. **43**(8): 1303–1333.
13. Campbell, JE, RP Thompson, J Dean and TW Clyne, Experimental and computational issues for automated extraction of plasticity parameters from spherical indentation. *Mechanics of Materials*, 2018. **124**: 118–131.
14. Kang, SK, JY Kim, CP Park, HU Kim and D Kwon, Conventional Vickers and true instrumented indentation hardness determined by instrumented indentation tests. *Journal of Materials Research*, 2010. **25**(2): 337–343.
15. Hernot, X, C Moussa and O Bartier, Study of the concept of representative strain and constraint factor introduced by Vickers indentation. *Mechanics of Materials*, 2014. **68**: 1–14.
16. Tiryakioglu, M and JS Robinson, On the representative strain in Vickers hardness testing of 7010 aluminum alloy. *Materials Science and Engineering: A Structural Materials: Properties, Microstructure and Processing*, 2015. **641**: 231–236.
17. Giannakopoulos, AE, PL Larsson and R Vestergaard, Analysis of Vickers indentation. *International Journal of Solids and Structures*, 1994. **31**(19): 2679–2708.
18. Alcala, J, AC Barone and M Anglada, The influence of plastic hardening on surface deformation modes around Vickers and spherical indents. *Acta Materialia*, 2000. **48**(13): 3451–3464.
19. Antunes, JM, LF Menezes and JV Fernandes, Three-dimensional numerical simulation of Vickers indentation tests. *International Journal of Solids and Structures*, 2006. **43**(3–4): 784–806.
20. Sakharova, NA, J Fernandes, JM Antunes and MC Oliveira, Comparison between Berkovich, Vickers and conical indentation tests: a three-dimensional numerical simulation study. *International Journal of Solids and Structures*, 2009. **46**(5): 1095–1104.
21. Weiss, HJ, On deriving Vickers hardness from penetration depth. *Physica Status Solidi A: Applied Research*, 1987. **99**(2): 491–501.

22. Atkinson, M, Further analysis of the size effect in indentation hardness tests of some metals. *Journal of Materials Research*, 1995. **10**(11): 2908–2915.
23. Wheeler, JM, RA Oliver and TW Clyne, AFM observation of diamond indenters after oxidation at elevated temperatures. *Diamond & Related Materials*, 2010. **19**: 1348–1353.
24. Haddow, JB, A study of Knoop hardness test. *Journal of Basic Engineering*, 1966. **88**(3): 682–&.
25. Giannakopoulos, AE and T Zisis, Analysis of Knoop indentation. *International Journal of Solids and Structures*, 2011. **48**(1): 175–190.
26. Tsui, TY, WC Oliver and GM Pharr, Influences of stress on the measured mechanical properties using nanoindentation: part I. Experimental studies in an aluminium alloy. *Journal of Materials Research*, 1996. **11**(5): 752–759.
27. Suresh, S and AE Giannakopoulos, A new method for estimating residual stresses by instrumented sharp indentation. *Acta Materialia*, 1998. **46**(16): 5755–5767.
28. Huber, N and J Heerens, On the effect of a general residual stress state on indentation and hardness testing. *Acta Materialia*, 2008. **56**(20): 6205–6213.
29. Rickhey, F, JH Lee and H Lee, A contact size-independent approach to the estimation of biaxial residual stresses by Knoop indentation. *Materials & Design*, 2015. **84**: 300–312.
30. Kim, YC, HJ Ahn, D Kwon and JY Kim, Modeling and experimental verification for non-equibiaxial residual stress evaluated by Knoop indentations. *Metals and Materials International*, 2016. **22**(1): 12–19.
31. Hosseinzadeh, AR and AH Mahmoudi, An approach for Knoop and Vickers indentations to measure equi-biaxial residual stresses and material properties: a comprehensive comparison. *Mechanics of Materials*, 2019. **134**: 153–164.
32. Aydin, A and A Basu, The Schmidt hammer in rock material characterization. *Engineering Geology*, 2005. **81**(1): 1–14.
33. Altindag, R and A Guney, ISRM suggested method for determining the Shore hardness value for rock. *International Journal of Rock Mechanics and Mining Sciences*, 2006. **43**(1): 19–22.
34. Kovler, K, FZ Wang and B Muravin, Testing of concrete by rebound method: Leeb versus Schmidt hammers. *Materials and Structures*, 2018. **51**(5).
35. Brencich, A and F Campeggio, Leeb hardness for yielding stress assessment of steel bars in existing reinforced structures. *Construction and Building Materials*, 2019. **227**.
36. Tabor, D, Mohs hardness scale – a physical interpretation. *Proceedings of the Physical Society of London Section B*, 1954. **67**(411): 249–257.
37. Gerberich, WW, R Ballarini, ED Hintsala, M Mishra, JF Molinari and I Szlufarska, Toward demystifying the Mohs hardness scale. *Journal of the American Ceramic Society*, 2015. **98**(9): 2681–2688.
38. Williams, JA, Analytical models of scratch hardness. *Tribology International*, 1996. **29**(8): 675–694.
39. Useinov, AS and SS Useinov, Scratch hardness evaluation with in-situ pile-up effect estimation. *Philosophical Magazine*, 2012. **92**(25–27): 3188–3198.
40. Kareer, A, XD Hou, NM Jennett and SV Hainsworth, The existence of a lateral size effect and the relationship between indentation and scratch hardness in copper. *Philosophical Magazine*, 2016. **96**(32–34): 3396–3413.
41. Narayanaswamy, B, A Ghaderi, P Hodgson, P Cizek, Q Chao, M Saf and H Beladi, Abrasive wear resistance of ferrous microstructures with similar bulk hardness levels evaluated by a scratch-tester method. *Metallurgical and Materials Transactions A: Physical Metallurgy and Materials Science*, 2019. **50A**(10): 4839–4850.

8 Indentation Plastometry

Indentation plastometry is now emerging as a potentially valuable addition to the range of testing techniques in widespread use. In many ways, it incorporates an amalgamation of the convenience and ease of usage offered by hardness testing with the more rigorous and meaningful outcomes expected of tensile testing. The indentation procedure itself is very similar to that of hardness testing, except that the loads required are higher than those used in most types of hardness test. The major difference is that the experimental data extracted are much more comprehensive, either in the form of a load–displacement plot or as a residual indent profile (with the latter offering several advantages). However, these experimental data only become useful if they can be processed so as to obtain a (true) stress–strain relationship, which can in turn be used to predict the (nominal) stress–strain curve of a conventional tensile test, including the strength (UTS) and the post-necking and rupture characteristics. This can only be done in a reliable way via iterative FEM simulation of the indentation process, but commercial packages in which this capability is integrated with a test facility are now becoming available.

8.1 Introduction to Indentation Plastometry

While hardness testing gives only a semi-quantitative indication of the resistance to plastic deformation, the outcome of an indentation operation – for example, the exact size and shape of the residual indent – does depend in a sensitive way on the (true) stress–strain relationship of the material, potentially over a large range of plastic strain. Hardness testing, which is usually based on just measuring a single parameter (indent depth or lateral size), extracts only a minute fraction of the information incorporated into this residual profile. There are also procedures for obtaining hardness values from load–displacement data acquired during indentation (using "nanoindenters" – see Chapter 9). These data, like those of the complete residual profile, are certainly much richer in information than single indent size values. However, the focus on a hardness value, usually defined as the contact stress under peak load, has meant that the outcome is still very limited in terms of characterizing the plastic deformation of the material (as explained in Chapter 7).

This is a classical challenge of the "*inverse*" type. If the stress–strain relationship is known, then the response during indentation (including the load–displacement curve and the residual indent shape) can readily be predicted (using FEM). However, inferring the stress–strain curve from those outcomes is far from simple. Broadly, there are two approaches to the problem. The conceptually simpler (and more rigorous) one is to repeatedly carry out the FEM simulation [1–13], using a constitutive law (§3.2.1) to represent the (von Mises) stress–strain relationship and systematically changing the values of the parameters in that law until optimum agreement is reached between measured and predicted outcomes (load–displacement plot and/or residual indent profile). While conceptually simple, several challenges and difficulties arise. These include the danger of a lack of "***uniqueness***" – i.e. the possibility of different combinations of parameter values (i.e. different stress–strain curves) giving effectively identical outcomes. There is also the question of how to ***converge*** (move in parameter space) on the "solution" in an efficient way. It is in any event likely that a large number of FEM simulations will be needed. While FEM capabilities are certainly more powerful and ubiquitous than in the past, this still constitutes a significant barrier to quick and easy implementation of the approach. Furthermore, there are several important issues related to the details of the experimental measurements and to the formulation of the FEM model (***mesh*** selection, specification of ***boundary conditions*** etc. – see §8.3.2).

The alternative methodology is to process the outcome data in some analytical way so as to obtain a "best-fit" stress–strain curve directly, without the need for any FEM simulation or iterative calculations. This obviously has major attractions. Even if the analytical operations are relatively complex, the computation time is likely to be negligible. Also, it may be unnecessary to limit the resultant stress–strain relationship to one conforming to a particular constitutive law – it may be possible, for example, to somehow convert the load–displacement data directly to a stress–strain relationship. In fact, many claims have been made suggesting that such operations are viable, often involving some kind of "correction" or "calibration" factor that is specific to the type of material. There have certainly been many proposals of this type [4, 14–22], often displaying considerable ingenuity and imagination.

The difficulty with this approach arises from the inherent complexity of the stress and strain fields created during indentation (compared with those during uniaxial loading). This complexity is partly due to the continuing expansion of the volume of material in which plastic deformation is taking place, which is inherent in any indentation procedure. There is simply no possibility of capturing this complexity in an analytical model. In practice, any such approach must involve empirical formulations and these are very unlikely to have any universal applicability. Despite the attractions, and the many attempts, any such approach is likely to suffer from severe limitations. It has become clear, after much investigation, that only the iterative FEM simulation approach offers potential for the development of a tractable and universally applicable tool. It is now becoming known as ***indentation plastometry***.

8.2 Experimental Issues

8.2.1 Indenter Shape and Size

There are strong incentives to use ***spherical indenters***. These advantages have been highlighted a number of times [10–12, 19, 23]. One motivation is that a sphere is much less prone to becoming *damaged* than are shapes having edges or points, and is also easier to *specify* and *manufacture*. Spheres (of WC-based cermets, with hardness and stiffness values high enough for most purposes), having diameters in the preferred range of about 1–4 mm (see below), are cheap and readily obtained. In general, they never need replacing. There is also reduced risk with spheres of encountering the computational problems that are often associated with simulation of behavior in regions of high local curvature (edges or points). Finally, at least with (approximately) isotropic materials, a spherical indenter allows the FEM modeling to be radially symmetric (2-D), which is not possible with most shaped indenters. The potential need for very large numbers of iterative FEM runs makes this a more significant issue than it would otherwise be.

As emphasized elsewhere (§5.1.3, §7.2.1, §7.2.3 and §7.3.1), it is important, when the objective is to extract bulk properties, to indent on a suitable *scale*, while retaining the key advantages of being able to test *small, flat samples*, to carry out *point-to-point mapping* of properties and to be an essentially *non-destructive* procedure. In particular, the volume being interrogated must have a (stress–strain) response that is representative of the bulk. It is on this *meso-scale* (creating indents large enough for representative material response, but small enough to allow small samples and mapping) that this type of measurement needs to be focussed.

The minimum indent size for representative response depends on microstructure, but in most cases it will require deformation of an *assembly of grains* – at least about a dozen and preferably more. Only when such an assembly is being deformed is it possible to capture the influence, not only of the crystallographic texture of the material, but also of the way that cooperative deformation of neighboring grains takes place. This is likely to be affected, not only by texture, but also by factors such as the ease of grain boundary sliding. Simply taking the average of the load–displacement responses, or indent profiles, from indents made in a large number of individual grains will not even approximately capture the bulk response. (The same arguments apply to carrying out conventional uniaxial tests on a set of single crystal samples having orientations representative of the texture of a polycrystal – see §4.3.1.) A crude rule of thumb might thus be that, viewed on the free surface, the indent should straddle at least "several" grains.

Of course, the corresponding minimum indent diameter might range from below 1 μm to above 1 mm, but it will certainly be small enough in most cases to offer the attractions outlined above. In practice, *grain sizes* of around 100 μm or more are common. In general, therefore, *indent diameters* (at the surface) should be at least a *few hundred microns*. (It may also be noted in this context that a relatively coarse scale of indentation minimizes the problems associated with surface roughness, oxide films, contamination etc. – see §8.2.3.) If the objective is to have a physical set-up that can be used across the range of likely sample microstructures, then an *indenter diameter* of the order of *1–2 mm* is likely to be appropriate. Much of the work described in this chapter relates to results obtained with an *indenter radius of 1 mm*.

It has long been recognized [24–29] that there may be advantages in obtaining more comprehensive sets of experimental data. For example, doing repeat runs with indenters having different shapes has often been proposed, and indeed it is logical that this could be helpful, since the way that the stress–strain curve influences the indentation outcomes will be different with different indenter shapes. On the other hand, as outlined above, a sphere offers important advantages over all other shapes, so this is not an attractive option in practice.

It has occasionally been suggested that using more than one indenter size may also be helpful. However, this is unlikely to create benefits (for bulk samples), since the stress and strain fields beneath an indenter are *scale-independent*. For example, the fields created by penetration of a sphere to a depth corresponding to, say, 10% of its radius are identical for indenter radii of, say, 10 μm and 10 mm. The absolute value of the load at this point will be 10^6 greater for the latter case, while the penetration will be 10^3 greater, but the information being provided about the stress–strain response of the material is the same, provided the volume being interrogated is in both cases large enough to be representative of the bulk response.

There is, however, a final point to note in this context, which is that, if the focus is on residual indent shape, there may be potential benefits from measuring it at more than one penetration depth. This is not, of course, relevant if the experimental data are those of load–displacement curves, in which case information is being acquired throughout the test.

8.2.2 Penetration Depth, Plastic Strain Range and Load Requirements

It is clear that the indenter *penetration depth*, δ, expressed as a ratio to the indenter radius, R, is an important experimental variable. It might be imagined that, while the load needed to penetrate to a given δ/R, and the stresses in the material, would depend strongly on the material (hardness), the strains would not. In fact, this is not really true, since materials with different work hardening characteristics tend to exhibit significantly different plastic strain fields (for a given δ/R). Furthermore, even if the peak strain created in the sample is, say, 40%, the indentation response will be considerably more sensitive to lower strain regions of the stress–strain curve, where most of the plastic deformation takes place.

There is also the issue of the range of strain over which a representation of the stress–strain relationship is required, or is likely to be reliable. For example, the main purpose of obtaining a (true) stress–strain curve (values of the parameters in a constitutive law) will often be to use it for predicting the outcome of a conventional tensile test (usually in the form of a nominal stress–strain plot). Prediction of the UTS (commonly corresponding to the onset of necking) only requires the law to be reliable up to that strain level, which is sometimes just a few % and rarely more than a few tens of %. If there is also interest in simulation of the *post-necking behavior*, however, then that might involve strains in the neck reaching values of the order of 50–100%. That is a regime in which conventional (uniaxial) testing cannot normally be used to explore stress–strain relationships, but it is often possible to carry out indentation tests in which such strain levels are created in a controlled way. It is in any event clear that an indentation test must create plastic strains of at least up to *a few tens of % –*

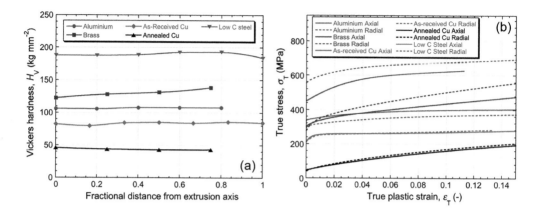

Fig. 8.1 Experimental data [12] for a set of five materials (in the form of extruded rods), showing (a) Vickers hardness numbers, at different radial locations in transverse sections and (b) stress–strain plots, from compressive loading in axial and radial directions.

otherwise the outcome will be relatively insensitive to the stress–strain relationship in the strain range of probable interest. This is explored in more detail below.

Figure 8.1 shows experimental data [12] from extruded rods of five different materials. With material manufactured in this way, there is always the possibility of strong crystallographic texture (and consequent **anisotropy** – see §8.4.2), and also of **inhomogeneity** – i.e. spatial (radial) variations in microstructure and properties. Figure 8.1(a), which gives Vickers hardness numbers from indents at different radial locations, confirms that any such spatial variations were not significant for these materials. However, Fig. 8.1(b), which shows stress–strain curves obtained via compressive loading in both axial and radial directions, indicates that three of the materials exhibited significant anisotropy, while the other two (i.e. the Cu-based ones) were more or less **isotropic**. It is also clear that these materials cover a fairly wide range of plasticity characteristics. Incidentally, it can be seen that these H_V values, if expressed as a stress (multiplied by g) and then converted to a yield stress (divided by a factor of 3, as in Eqn. (7.5)), give a value that corresponds (approximately) to a "*flow stress*" averaged over a strain range up to about 10–15%, which is consistent with points made in §7.2.3.

Figure 8.2 shows load–displacement plots [12] obtained during indentation of these five materials, using a spherical indenter of diameter 2 mm. (These are all averages from a number of indents, although there was very little scatter: virtually identical curves were obtained when indenting in different directions [12].) An immediate point to note here relates to the magnitude of the load needed to generate **penetration ratios** of the order of 20–40%, with this size of indenter. While none of these materials are very hard, these loads are well into the **kN range**. This is far beyond the limit of "nanoindenters," which are the machines most commonly available with accurate **depth-sensing** capabilities (§9.2.1). A different type of machine is needed for

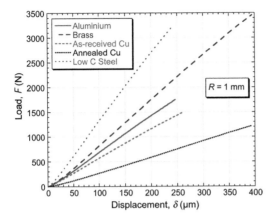

Fig. 8.2 Experimental load–displacement data [12] from indentation (with a 2 mm diameter sphere) on transverse sections, for the set of five materials.

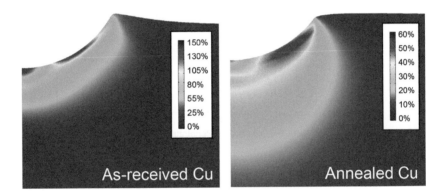

Fig. 8.3 Predicted [12] plastic (von Mises) strain fields after indentation to a penetration ratio (δ/R) of 40%, for as-received and annealed Cu.

indentation plastometry, combining a relatively high load capacity with a capability for accurate measurement (of load–displacement curves and/or residual indent profiles).

Furthermore, as outlined above, the **distribution of plastic strain** created during indentation is important in terms of the sensitivity of the outcome to the required objective (true stress–strain curve over an appropriate range of plastic strain, which might be up to at least several tens of %). This is dependent, not only on the penetration ratio, but also on the material response. For example, the two strain fields [12] in Fig. 8.3 highlight this dependence, showing that the strains are higher in the as-received Cu, due to its lower **work hardening rate**, compared with that of the annealed Cu. (This also promotes much more **pile-up** with the as-received Cu.)

The outcome of a systematic study [12] into this issue can be seen in Fig. 8.4, which shows **distributions of plastic work** done in different **ranges of plastic strain**,

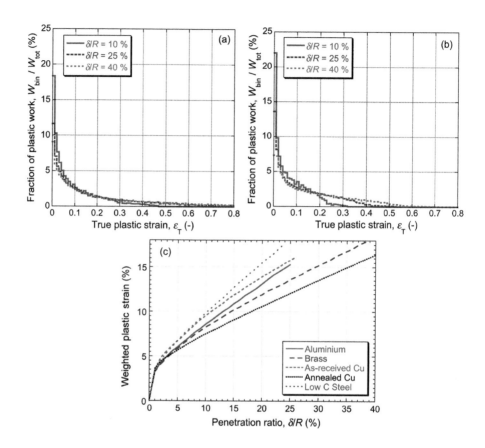

Fig. 8.4 Fractions of the total plastic work done in different ranges of strain [12] during penetration to three different penetration ratios into: (a) as-received Cu and (b) annealed Cu, plus (c) average plastic strains for all five materials, as a function of penetration ratio.

for these two materials. Also shown are the average plastic strain levels (weighted for the work done), as a function of penetration ratio, for all five materials. It's difficult to make universally applicable statements regarding the optimum level for this average, but it's probably desirable for it to be at least about 10%, leading to a requirement for a penetration ratio of around 20%. This is only an approximate figure, but it can nevertheless be regarded as a useful guideline for the design and operation of commercial machines. With a very hard metal, and an indenter of 2 mm diameter, the *load requirement* for this penetration ratio is about 6–8 kN.

8.2.3 Sample Preparation

One of the attractions of indentation plastometry is that the only requirement regarding sample preparation is that the region concerned should be *flat* and *relatively smooth*. The issue of *surface roughness*, and the effects of *contamination*, *oxide films* etc., are

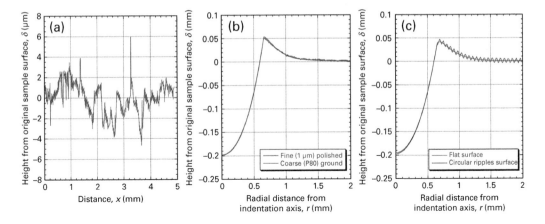

Fig. 8.5 Effect of surface roughness on indent profiles (for an indenter of radius 1 mm penetrating into an annealed C250 maraging steel sample (having Ludwik–Hollomon parameter values of $\sigma_Y = 1000$ MPa, $K = 500$ MPa and $n = 0.7$, with $\mu = 0.3$): (a) surface profile after grinding with P80 grit paper, (b) experimental indent profiles for fine polished and coarse ground surfaces and (c) FEM modeled indent profile for a flat and "rippled" surface.

obviously relevant here. It would normally be expected that, if the ratio of the penetration depth to the scale of such roughness and contamination is large, then their effects can safely be ignored. A minimum magnitude for this ratio cannot be accurately specified, but a value of, say, 20 should be sufficient. If $R \sim 1$ mm and $\delta/R \sim 20\%$, then $\delta \sim 200$ μm, so that any roughness/oxide film thickness less than about 10 μm should not present any significant problems. This is actually a relatively coarse surface finish, and it would certainly be an unusually thick oxide film. Of course, a very rough surface would be unacceptable, but the necessary *surface preparation* can be fairly *crude and simple*.

This is illustrated by the plots in Fig. 8.5, which show measured and predicted indent profiles created on well-polished and coarsely ground surfaces – P80 paper is impregnated with grit of about 200 μm diameter. It can be seen (in Fig. 8.5(b)) that, even with this rather coarse surface preparation (giving an R_a value of about 1 μm and an R_z value of about 7.5 μm), the experimental residual indent profile is virtually indistinguishable from that of a well-polished surface. This is consistent with the modeled profiles shown in Fig. 8.5(c), for which a rough surface was simulated by introducing a sinusoidal topography, with an amplitude (peak to trough) of 10 μm and a wavelength of 60 μm, creating circular "ripples" centered on the indentation axis. This is clearly a very "rough" surface, but again the resultant indent profile is very similar to that from a flat surface, particularly if it is smoothed beyond the rim of the "crater." Load–displacement plots are also fairly insensitive to roughness, although the initial part (where deformation of asperities is taking place more easily than would penetration into the bulk) can be affected in terms of the zero position of the plot. Of course, the key issue is the scale of the roughness relative to the dimensions of the indent, so fine scale indentation is much more sensitive to roughness (and oxide films etc.) – see §9.2.1.

However, it should be noted that grinding and polishing operations carried out on a sample prior to testing may leave a *near-surface region* with different properties from that of the bulk. This is most likely with a soft material, for which the production of a relatively thick *work-hardened layer* is a distinct possibility. Again, use of relatively large indenters and deep penetration reduces the sensitivity of the outcome to the presence of such layers. However, even with a penetration depth of, say, 200 μm, a hardened layer that is, say, 20 μm thick could affect the outcome. The upshot is that care is needed with *very soft materials* – those with a yield stress below about 30 MPa – and that *prolonged fine scale polishing* during the final stages may be needed with such materials.

These issues of surface roughness/oxide layers etc. are, of course, different from that of whether indentation can be carried out on a surface that is not actually flat, but is systematically *curved* – for example, the surface of a tubular component, such as a pipeline. This is often more relevant to *in situ* testing, as opposed to use of a bench-top machine in a laboratory. This is potentially an issue whether the focus is on load–displacement data or on residual indent shape. For the case of residual indent shape, a minimum curvature likely to have a significant effect can be estimated from the probable length of a scan and the resolution required. With $R \sim 1$ mm and $\delta/R \sim 20\%$, the indent diameter is about 600 μm and the total scan length is likely to be about 4 mm (fully capturing any pile-up effect). If a (relatively high) surface roughness of 10 μm is taken to be a maximum acceptable height difference over a distance of 4 mm, this corresponds to a curvature of about 1 m^{-1} (radius of curvature 1 m) – the geometrical construction for this calculation is shown in Fig. 8.6. This could therefore be a potential source of error even with a large diameter pipe. It can, however, be tackled in several ways. One is to somehow create a *small, flat region* on the surface of the pipe. Another is to scan over the region *before and after indentation*, and subtract one profile from the other (rather than implicitly assuming that the initial profile is flat). A third is to incorporate the curvature into the FEM model.

Finally, it's possible that the surface could be *undulated* in some way. This could occur on various scales, from that of a typical surface roughness (with a "wavelength" of the order of, say, 10 μm) to a much larger (mm or cm) scale of variation – arising, for example, from a *chemical etching* process. Indentation plastometry (producing an

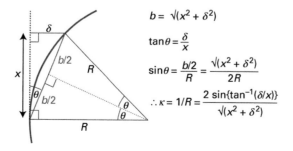

Fig. 8.6 Geometrical relationship between curvature and deflection.

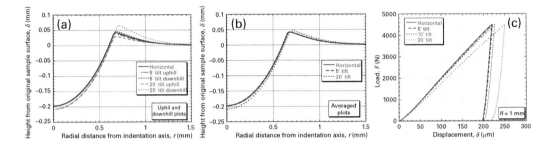

Fig. 8.7 FEM model predictions for indentation (with a sphere of radius 1 mm), into a material with Ludwik–Hollomon parameter values of: $\sigma_Y = 1000$ MPa, $K = 500$ MPa and $n = 0.7$, with $\mu = 0.3$, illustrating the effect of the angle θ between the indentation axis and the normal to the surface of the sample. These plots show (a) residual indent profiles, for a load of 4.5 kN, (b) averaged values of these and (c) load–displacement plots.

indent diameter of the order of 1 mm) is insensitive to the former, but could in principle be affected by the latter. This is really a question of whether outcomes are affected by an *inclination* between the axis of indentation and the normal to the surface being indented. Of course, this could arise, not only from an undulating surface, but also from the top and bottom surfaces of the sample not being perfectly parallel. It probably has to be accepted that sample surfaces inclined at, say, 2–3° to the horizontal are likely to be fairly commonplace. However, the FEM outcomes shown in Fig. 8.7 indicate that this is unlikely to be a concern. This figure shows, for a typical (true) stress–strain relationship (captured via a set of Ludwik–Hollomon parameter values), that such inclinations have a negligible effect on both load–displacement and residual indent data, with the latter being particularly insensitive to *tilt effects*. Of course, this is not true for very large inclinations (>~15°), but such cases would be unusual (for relatively coarse indentation).

It may also be noted that other effects related to *sample geometry* could arise. For example, if indentation is carried out close to a *free lateral surface* (an "*edge*") – whether this is actually free or is embedded in a (much more compliant) mounting medium, then this could affect the outcome. If such an edge is sufficiently close to the plastic deformation field, then the *radial symmetry* could be lost. How this can be quantified in an FEM model in terms of relative distances etc. is an issue concerned with mesh specification, as is the depth and lateral extent of the meshed domain needed to satisfy the condition of a "semi-infinite" system. This is addressed below in §8.3.2.

8.2.4 Experimentally Measured Outcomes

It should first be noted that there are certain requirements concerning the accuracy of the experimental data. Convergence on the "correct" stress–strain relationship will not be possible if the "target" data are noisy or unreliable. There are two main options regarding the type of target data to be obtained. Much of the early work on indentation plastometry was based on *load–displacement data*. This was a natural progression from the proliferation of "nanoindenters," with which the main activity has for some time been focussed on manipulation of such data (some of which has been oriented towards measurement of hardness) – see §9.3.2. Indeed, much of the initial indentation plastometry work was carried out using nanoindenters, despite the many problems that arise with very shallow penetration. The move towards testing on a coarser scale, using test set-ups with higher load capability, has solved some of these problems, relaxing the requirements in terms of the absolute resolution of displacement measurements, as well as reducing the errors likely to arise from surface roughness, oxide layers etc. (and improving the chances of the deformed volume being large enough to represent the bulk response). Nevertheless, noise in the displacement data is a potential issue. Moreover, depending on exactly how the displacement is being measured, there may be concerns about effects of the *compliance* [30–33] of the *loading train*: this contribution to the measured displacement must be subtracted, since it cannot be included in universal software for modeling of the indentation process.

There has therefore been increasing interest [13, 33–36] in using the ***residual indent profile*** as the main experimental outcome. This offers important advantages. It means that no measurements at all need to be made during the actual test, apart from noting the final value of the load. There need also be no concerns about the compliance of the loading train. These advantages are significant and are likely to be particularly relevant to *in situ* field testing of components. It naturally requires a procedure for (accurate and reasonably rapid) profile measurement. However, this also is easier when the scale of the indent is relatively coarse and there are several established (mechanical and optical) *profilometry techniques* that are likely to give the required resolution and ease of measurement.

On the issue of exactly how the profilometry is carried out, it is worth noting that there is an option for use of a *replica technique*. These are based on the injection of some kind of polymeric material into the region of the indent, followed by removal of the solidified replica and later measurement of its topography. Such procedures have a long history [37], mainly in the context of surface roughness and examination of other topographic features. It offers potential advantages for *in situ* work, when it may be difficult to use either optical or mechanical profilometry techniques. It may also be noted that there is often a particular difficulty when applying optical techniques to indents in metallic samples, which arises from the *high reflectivity* of the surface. This often causes problems in the vicinity of the "rim" of the indent, where the exact shape of the profile needs to be accurately captured. Of course, a replica is likely to be much less reflective, so this kind of problem does not normally arise.

8.3 FEM Simulation Issues

8.3.1 Representation of Plasticity Characteristics

For any approach involving iterative FEM simulation of a deformation process, the stress–strain relationship (material plasticity response) must be characterized via a (small) set of parameter values. Of course, for a single simulation, it would be possible to use an arbitrary set of **stress–strain data pairs**, but when the objective is to infer the optimal relationship consistent with obtaining a particular outcome, this leaves too many degrees of freedom for tractable convergence, so a functional form (involving a relatively small number of parameters) is required. Several expressions have been proposed, although two of them (the Ludwik–Hollomon and Voce equations – see §3.2.1) are in more common use than the others. It should be emphasized that these are all purely empirical relationships. There have been many efforts to rationalize stress–strain curves in terms of microstructural features (and their effect on dislocation mobility, which is at least the primary factor determining the ease of plastic deformation in metals). However, the concept of predicting stress–strain curves on the basis of identifiable microstructural features has not proved to be workable and the formulations in use are simply based on empirical fitting to experimental data.

A further point worthy of note concerns the possibility of **asymmetry** between (uniaxial) stress–strain curves obtained in **tension and in compression**. Any difference between the two is in principle indicative of a dependence of yielding (and subsequent progression of plastic straining) on the **hydrostatic component** of the stress state. In general, as described in §6.3.1, while differences are sometimes observed, they are normally due only to experimental difficulties – often associated with friction and consequent barreling in compression or necking in tension. Genuine asymmetry is only rarely observed, at least for metals: the probable origins in such cases are outlined in §6.3.1. This insensitivity of the plasticity response to the hydrostatic stress (§3.1.1 and §3.2.1) is consistent with the incompressibility of metals during plastic deformation and the main mechanisms of plastic deformation. The shear stress needed to cause dislocation glide (and also deformation twinning, which is significant in some cases) is independent of the hydrostatic stress and so the uniaxial stress–strain relationship is expected to apply under any stress state, provided attention is focussed on the **deviatoric (von Mises) stresses and strains**. This is implicit in virtually all FEM modeling of the plastic deformation of metals.

In addition to the Ludwik–Hollomon and Voce laws (Eqns. (3.13) and (3.14)), several other stress–strain relationships have been proposed, some aimed primarily at non-metallic systems such as polymers and rubbers. Some, such as that of **Ramberg–Osgood**, are oriented towards cyclic loading and a focus on the transition between elastic and plastic deformation [38]. There are others, such as the **Ludwigson** relationship, as described, for example, in Samuel and Rodriguez [39], that are effectively combinations of the above two, but involve more parameters and hence are likely to slow down convergence considerably. While there could be a motivation in some circumstances for exploring a wider range of formulations than the above two, in general the stress–strain curves of most metallic materials can be captured reasonably well using at least one of these two equations.

8.3.2 Meshing, Boundary Conditions and Input Data

There are several choices that need to be made in setting up an FEM model for simulation of indentation. **Sensitivity analyses** need to be run in order to confirm that

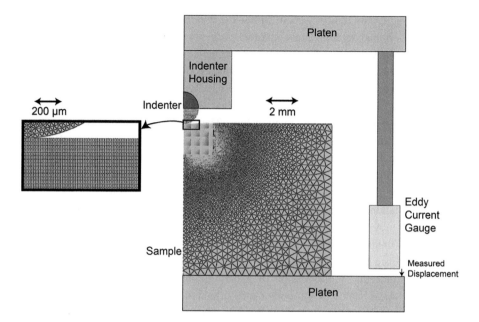

Fig. 8.8 Schematic representation of an indentation testing set-up, showing the mesh used in an FEM model [12].

the mesh is sufficiently fine to achieve **convergence**, **numerical stability** and **mesh-independent** results. This often requires **refinement** in regions where plastic strains are likely to be high. It's also important that the domain being modeled should be of a suitable size. In fact, there are many such issues that need to be taken into account when setting up an FEM model. There are excellent and comprehensive review articles available [40, 41], which cover them in the context of materials modeling.

In many indentation scenarios, an **axisymmetric** FEM model is appropriate. A typical mesh [12] is illustrated in Fig. 8.8. There are about 5,000 volume elements in this case, all **second-order quadrilateral** and/or **triangular**. The mesh is refined in regions of the sample close to the indenter, as shown. In this case, the complete sample was included in the simulation, with its lower surface rigidly fixed in place. In modeling the complete sample, contributions to the displacement caused by its elastic deformation, as well as plastic deformation, are fully captured. (Of course, this is only relevant when the focus is on load–displacement data.) In fact, the sample thickness is usually not important beyond a depth of about four indent diameters, since the axial stress – and hence the elastic strain – below that becomes negligible for typical cases. The lateral extent, beyond about four indent diameters, is also unimportant.

One area to be addressed is that of the effect of *friction* between indenter and sample [42–44]. This topic is covered in the context of uniaxial compressive testing in §6.2.2. The effect of interfacial friction is routinely simulated via a **coefficient of friction**, μ, such that sliding between the two surfaces requires a shear stress, τ, given by

$$\tau = \mu \sigma_n \tag{8.1}$$

where σ_n is the normal stress at the interface. The value of μ is expected to depend on the surface roughness of indenter and sample, and cannot really be predicted a priori. It is difficult, if not impossible, to directly measure a coefficient of friction under the conditions of an indentation test (during which new sample surface is being continually created). It's also difficult to vary the value experimentally in any systematic way – both indenter and sample are expected to be smooth and, since the contact pressure is high, lubricants are unlikely to remain in place – see §6.2.2. Modeling experience has shown [12] that the predicted behavior can exhibit some sensitivity to the value, particularly as the penetration ratio starts to become relatively large (>~10%), when the probability of interfacial sliding (near the periphery) often rises. (Of course, the value of μ is only of potential significance if there is in fact at least some sliding.) It's also worth noting that static and dynamic values may differ – i.e. it may drop slightly once sliding starts. There are difficulties in trying to include these degrees of freedom in the convergence operation, so some kind of averaged, universal value needs to be identified.

Relevant plots are shown in Figs. 8.9 and 8.10, which compare experimental and predicted residual indent profiles, with the modeled curves obtained using several different values of μ. These comparisons concern two Cu samples, similar to those of Fig. 8.3, one in the as-received (extruded bar) condition and the other being the same material after an annealing treatment (that stimulated recrystallization – see §4.2.4). They differ substantially in their work hardening characteristics, although it's also worth noting that (nominally similar) extruded copper can itself vary significantly in mechanical properties from batch to batch, mainly due to effects of variations in the

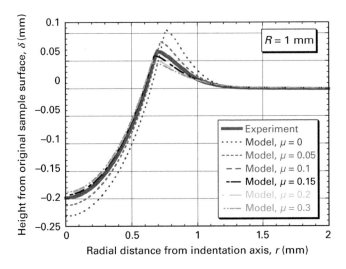

Fig. 8.9 Comparison, for as-received (extruded) copper bar, between experimental and FEM predicted residual indent profiles, with several different values of the friction coefficient. The predicted plots were obtained using the set of (Voce) plasticity parameter values giving best fit with a tensile stress–strain curve ($\sigma_Y = 355$ MPa, $\sigma_s = 355$ MPa and $\varepsilon_0 = 1$).

oxygen content and the extrusion conditions. Oxygen is an interstitial solute in copper and some details regarding its effects are provided in §4.3.3.

It should first be noted that the effect of the value of μ on the indent profile is clearly much stronger for material in the as-received state than after annealing. This is because the AR-Cu has very little capacity for work hardening. (As noted above, this may vary somewhat for such material, depending on the oxygen content and forming conditions, but this particular material doesn't really work harden at all, as can be seen from the fact that the best-fit values of yield stress and Voce saturation stress are the same.) Such material has a tendency for high local strains to be developed during indentation, mainly around the rim of the crater, with strong "*pile-up*" formation. This effect can be seen in Fig. 8.3(a). It can be seen in Fig. 8.9 that this is particularly pronounced if the value of μ is zero or very low, since the formation of pile-ups is facilitated by interfacial sliding.

It can also be seen in Fig. 8.9 that the best-fit value of μ is about 0.1. That this is a broadly appropriate value is in fact a rather general conclusion, although it can be seen most clearly with material like this that exhibits minimal work hardening. This value has been used in most of the simulations presented in this chapter, although the differences introduced if a larger value is used are relatively small. Furthermore, it is clear from Fig. 8.10 that, with materials that exhibit strong work hardening, which acts to inhibit pile-up (and favor "*sink-in*"), there is in any event much less interfacial sliding and the value of μ has very little effect.

Other input data required include the elastic constants (Young's modulus and Poisson ratio) of the sample material, plus those of the indenter. Both materials are assumed by default to be isotropic. Regarding the indenter itself, **anisotropy** is

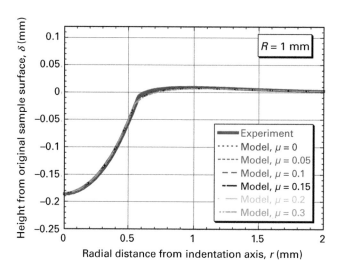

Fig. 8.10 Comparison, for annealed copper bar, between experimental and FEM predicted residual indent profiles, with several different values of the friction coefficient. The predicted plots were obtained using the set of (Voce) plasticity parameter values giving best fit with a tensile stress–strain curve ($\sigma_Y = 37$ MPa, $\sigma_s = 375$ MPa and $\varepsilon_0 = 0.19$).

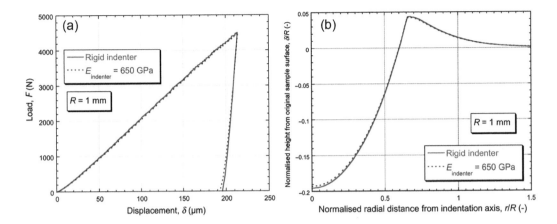

Fig. 8.11 Illustration of the predicted effect of taking a spherical indenter to be infinitely rigid, for indenting of an annealed C250 maraging steel (Ludwik–Hollomon parameter values: $\sigma_Y = 1000$ MPa, $K = 500$ MPa and $n = 0.7$), on (a) the load–displacement plot and (b) the corresponding residual indent profile.

certainly possible, particularly if it is a single crystal (as many "nanoindenter tips" are). For example, the elastic anisotropy of diamond [45] is potentially significant – about 10% range in Young's modulus, and considerably greater variation in Poisson ratio. However, indentation plastometry is carried out with relatively large spherical indenters and these are normally polycrystals (commonly cermets) that are effectively isotropic. (Of course, the sample could exhibit anisotropy – see §8.4.2.)

One question of interest is whether the indenter can be treated as a rigid body (removing the need to mesh it or to simulate stresses and strains within it). In fact, this assumption is often justifiable. For example, Fig. 8.11 shows the predicted effect (on load–displacement plot and residual indent profile) of treating a spherical (cermet) indenter as rigid (compared with using the correct elastic constants). This has been done for a relatively hard sample (of maraging steel), since differences are expected to be more significant in such cases. Nevertheless, it can be seen that the effect is small. Since it's not entirely negligible, there is an argument for including the indenter in the simulation – it may in any event be of interest to monitor the stress levels in it during indentation – but for most cases the key outcomes will be little affected.

Simulation runs are normally carried out under ***displacement control***, with the main outputs being predicted loads at a series of (~50–100) specified displacement values and/or the final residual indent shape (i.e. the height at a series of radial distances). One point to note here is that, while the profile should be monitored well beyond the "rim" of the indent, so as to capture fully the topography of any pile-up region, the weighting given to obtaining good agreement should be greater in the vicinity of the rim than elsewhere (since the shape in this region has a high sensitivity to the details of the stress–strain relationship – particularly the work hardening characteristics). Of course, the modeling also allows prediction of the stress and strain fields at any point during the process.

8.3.3 Characterization of Misfit ("Goodness-of-Fit")

Convergence on the *"target"* (i.e. experimental) outcome, whether that is the load–displacement plot or the residual indent profile, requires quantification of the *goodness-of-fit* for each modeled case. There are several options in such a scenario, but the most popular involves *minimization* of the **sum of the squares of the residuals**. This is a *"misfit" parameter*, which needs to be minimized.

For example, if the focus is residual indent shape, this sum, S, can be expressed:

$$S = \sum_{i=1}^{N} (\delta_{i,M} - \delta_{i,E})^2 \tag{8.2}$$

where $\delta_{i,M}$ is the ith value of the modeled height difference (predicted by FEM) and $\delta_{i,E}$ is the corresponding experimental (target) value, while N is the total number of radial locations being considered. The value employed for N can be varied, but it might typically be around 50. Since S is dimensional, it has units and its magnitude cannot be used to give a universal indication of the quality of the fit. For this purpose, therefore, it's common to use the parameter S_{red}, a *"reduced sum of squares,"* defined by

$$S_{red} = \frac{\sum_{i=1}^{N} (\delta_{i,M} - \delta_{i,E})^2}{N \delta_{av,E}^2} \tag{8.3}$$

where $\delta_{av,E}$ is the numerical average of the highest and lowest of the experimentally measured heights. Usually, the value of N is not pre-determined, but is dependent on the radial range. This might extend to a position such that the height had returned to within 5 μm of the far-field level (i.e. the level of the original flat surface). If the interval between radial locations were 10 μm, so that, if the height reached a level within 5 μm of the original surface at a radial distance of, say, 1.6 mm, then the value of N would be 160. Of course, such details are subject to optimization in various ways.

The parameter S_{red} is thus a positive dimensionless number, with a value that ranges upwards from 0 (corresponding to perfect fit). As a generalization, modeling that captures the material plasticity response reasonably well should lead to a solution (set of parameter values) for which S_{red} is less than, say, 10^{-3}. This effectively constitutes a **health check** on the solution – if, for example, no solution can be found giving a value smaller than, say, 1%, then this suggests that there can only be limited confidence in the inferred set of values. This could be due to experimental deficiencies and/or an inability to capture the behavior well with the constitutive law being used.

8.3.4 Convergence Algorithms

The best-fit set of values is obtained via iterative improvement, using a search algorithm. Again, there are many options [46–50], but a popular and illustrative one is the *Nelder–Mead simplex search* [51]. For example, the implementation suggested by Gao and Han [52], built using the Scientific Python and Numeric Python packages [53, 54] can be used. This is briefly outlined below.

For a model with m parameters, searching is within an m-dimensional parameter space, within which a *simplex* is defined. This is a **polytope** with (m+1) vertices (i.e. a triangle in 2-D, a tetrahedron in 3-D etc.). Each vertex corresponds to a particular combination of all of the m parameters in the set and the simplex covers a range of values for all of these. These points can be expressed as vectors (first rank tensors) in parameter space, designated $x_1, x_2, \ldots x_{m+1}$, each of which consists of a set of m parameter values. After each iteration (new FEM simulation), the objective is to "improve" the simplex by replacing the worst vertex (i.e. the one with the highest value of S) with a better point. The search for this better point is along a line in parameter space defined by the worst point and the centroid of the rest of the simplex, which is the average position of the remaining points (after removal of the worst point). The steps involved in the algorithm are described in Appendix 8.1.

The algorithm is terminated once a specified convergence criterion has been met. This can be defined as a relative difference (commonly 10^{-4}) in S and/or x between successive iterations. The number of iterations to achieve convergence depends on a number of factors, in addition to this criterion specification. These mostly relate to the way that the misfit varies in parameter space, which in turn depends on several issues (including how well the stress–strain curve can be captured by the selected constitutive law). There may in some cases be a danger of converging on a local minimum. Difficulties can also arise from the presence of "plateau regions," where various parameter value combinations give very similar degrees of fit. It may be helpful to optimize the preset values in the algorithm – see Appendix 8.1.

Overall, the computational operation is tractable in most cases. The procedure could, however, be facilitated if the starting values for the simplex were in an appropriate region of parameter space – i.e. if the initial trial values are fairly close to the best solution set. One way of ensuring that these starting values are at least in the right part of parameter space would be to generate a pre-run matrix of simulations, and resultant load–displacement plots or residual indent profiles, and simply select the one giving the best fit (lowest S value). One drawback of this is that all such "pre-running" would need to be done for specified values of certain parameters that are not in the set being searched, notably the elastic constants of the material and the friction coefficient.

An example [12] of the progression of a convergence procedure is shown in Fig. 8.12. This usually requires only a relatively small number (a few tens) of iterations, although of course this does depend on the level of accuracy being sought. It can be seen how the movement through parameter space typically takes place. The starting values were selected in a fairly arbitrary way and, as can be seen, they were not very close to the optimized values in this case. If they happened to be either very close to, or to differ substantially from, the optimized set, then that would affect the efficiency of the convergence. However, there would not, in general, be very much difference between the number of iterations required in the two cases. On the other hand, there will always be a requirement for more than just a few iterations, even if the starting positions are quite close (obtained, for example, by scanning a pre-run set).

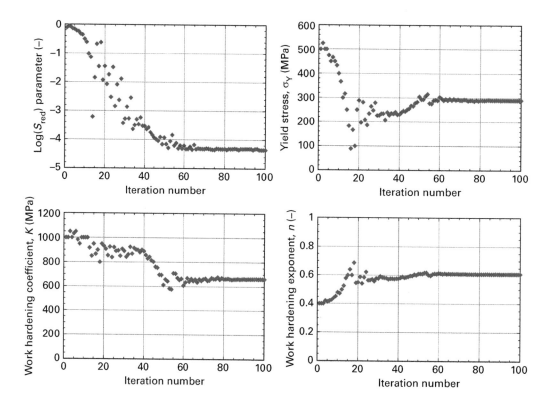

Fig. 8.12 Nelder–Mead convergence [12] on an optimal (Ludwik–Hollomon) parameter set, targeting a load–displacement plot from indentation of a brass sample, showing the evolution with iteration number of: (a) the goodness-of-fit parameter, S_{red}, (b) yield stress, (c) work hardening coefficient and (d) work hardening exponent.

8.4 Range of Indentation Plastometry Usage

8.4.1 Presentation of Results

The key outcome of an indentation plastometry experiment is likely to be a best-fit set of parameter values in a constitutive law. This can be used to obtain the true stress–true (plastic) strain relationship, in principle over any range of strain. In practice, there will also be information about the probable strain range over which it's likely to be reliable. Of course, this relationship constitutes fundamental information about how the material deforms plastically. However, the comparison that is most likely to be made is with the outcome of a conventional tensile test, which is often available only as a nominal stress–nominal strain plot (including a final rupture point). It is straightforward to use the indentation-derived true stress–true strain relationship in an FEM model of a tensile test, as described in §5.3.3, so as to obtain such a plot. This can even include a predicted rupture point, by simulating the final necking characteristics and using a fracture criterion (such as a critical level of plastic strain being reached in the

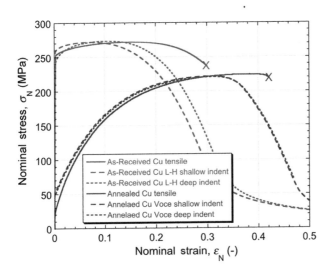

Fig. 8.13 Comparisons [13] between nominal stress–strain curves obtained directly by tensile testing and via measurement of residual indent profiles, with two different depths.

neck). This is how the plots shown in Fig. 5.10 were obtained, although, as noted in §5.3.3, this final portion of the curve is unlikely to resemble closely the experimental one, even if the tensile sample dimensions (which affect this final part) are correctly specified.

A typical example of such a comparison [13] is shown in Fig. 8.13, for two types of Cu-based material, similar to those referred to in Figs. 8.1–8.3. It can be seen that the agreement is good regarding the yielding, work hardening and UTS characteristics. In these plots, the FEM simulation of the tensile test did not include the correct sample dimensions, so it does not include any attempt to predict reliably the necking down part of the plot or the final rupture point. Of course, it's also possible to obtain modeled stress and strain fields, both during the indentation procedure and for the tensile testing. Among further FEM outcomes of potential interest is how the plastic work is distributed in terms of prior strain during the indentation. This can be analyzed using the algorithm described in Appendix 8.2, which was employed in creating Fig. 8.4.

8.4.2 Effects of Material Anisotropy and Inhomogeneity

Among factors of potential significance are **anisotropy** (different responses in different directions of testing) and **inhomogeneity** (different responses in different parts of the same sample). Of course, this also applies to other types of testing, including conventional uniaxial loading, but some differences are expected between that case and indentation. Firstly, indentation is expected to be more suitable for exploring **point-to-point variations** in properties – indeed, this is one of its key attractions. This can be taken down to a very fine scale, examining the responses of individual phases or thin surface coatings, although attention is focussed here on testing of volumes sufficiently large to ensure a representative "bulk" response. Nevertheless, the bulk response could vary significantly between different parts of a relatively large sample or component and indentation is attractive for exploring this.

Anisotropy is a slightly different issue. It can arise in several ways in different types of material, but the commonest cause, at least in metallic polycrystals, is *crystallographic texture*. Most metallic samples are textured, at least to some degree, and this will, in general, lead to both elastic and plastic anisotropy. Properties such as Young's modulus and yield stress often vary with direction by several tens of % in single crystals. While the variation in a polycrystal, even a strongly textured one, is expected to be less than this, differences of up to about 10% are quite common. These variations may be significant compared with the level of precision being sought in the inferred plasticity parameters. Some examples of this are apparent in Fig. 8.1, in which significant anisotropy can be seen between the uniaxial compression responses in axial and radial directions of (three of the five) extruded rod samples. Some basic information about relationships between texture and anisotropy, particularly for sheet metal, is provided in §5.4.2. More comprehensive information is, of course, available in the literature [55–58]. It may also be noted, however, that anisotropy can arise in other ways. For example, it can be quite pronounced in MMCs (metal matrix composites) containing aligned reinforcement of some sort – see below.

Conventional uniaxial testing will clearly provide information about the anisotropy of a material (provided samples can be obtained that are suitable for testing the material in different directions, which may be difficult in some cases). It is not, however, immediately clear whether the indentation response of an anisotropic material will be significantly different in different directions. Certainly both stress and strain fields are much more multi-axial than in conventional testing, particularly for relatively deep penetration. In fact, there is now a considerable body of evidence that virtually the same indentation response is obtained with the axis in different directions, even for (polycrystalline) samples that exhibit quite pronounced anisotropy. For example, the indentation responses of all of the materials in Fig. 8.1 were effectively identical in different directions [12].

On the other hand, the presence of (in-plane) anisotropy is often apparent in the form of a residual indent (created using a spherical indenter) that is not radially symmetric. An extreme example of this can be seen in Fig. 7.12(a), which shows the free surface of an indent in a superalloy single crystal. Of course, while pronounced effects of this type are only expected with an indenter of mm dimensions if the sample is a single crystal, or has a very large grain size, they commonly arise during "nanoindentation," when indent dimensions are often of the order a few μm or less and they are commonly located in a single grain. It may, however, be noted that most such indentation is done using "sharp" indenters, such that the indent is in any event not radially symmetric and the effects of (in-plane) anisotropy are less noticeable.

Nevertheless, there is potential for such anisotropy (in a polycrystal) to be detected, and possibly quantified, by careful study of the shape of the indent produced during

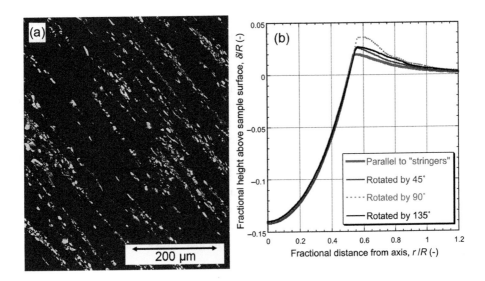

Fig. 8.14 (a) SEM micrograph of an Al MMC forging reinforced with about 7 vol% of the $Al_9Ni_{1.3}Fe_{0.7}$ intermetallic phase and (b) profiles across an indent in it, made with a 1 mm radius sphere and obtained by scanning in four different directions, relative to the alignment of the reinforcement.

indentation plastometry. This primarily relates to plastic, rather than elastic, responses, although elastic anisotropy is in any event expected to be relatively small in most cases. An example of the type of effect that can be observed is shown in Fig. 8.14. This relates to a material that is effectively an Al-based MMC, containing about 7 vol% of a (hard) Al-Fe-Ni intermetallic phase. (More detailed information about such materials can be found in MMC textbooks [59].) This "reinforcement" was aligned (during a forging process) into the "stringers" that can be seen in Fig. 8.14(a). This rendered the material (plastically) anisotropic, being harder in the direction of alignment. This effect is apparent in the residual indent profiles shown in Fig. 8.14(b), which were measured (mechanically) in directions parallel, transverse and at ±45° to the alignment direction. Since the indent is over 1 mm wide, and both the grain structure and the scale of the reinforcement distribution are much finer than this, these differences in the profile clearly reflect the "bulk" anisotropy of the material. Incidentally, this constitutes another advantage of using the residual indent profile to characterize the material, rather than the load–displacement plot, since the latter could not be analyzed to reveal information about anisotropy.

8.4.3 Measurement of Residual Stresses

It has long been clear that, during indentation, the presence of residual stress in the near-surface region of a sample can affect the way that plasticity develops and hence the details of the penetration characteristics (and the main measurable outcomes – i.e. the load–displacement plot and residual indent profile). All that needs to be done is to superimpose the residual stresses and those from the applied load. This point is made in §3.3.2. Of course, the stress field from the applied load is complex, so this will need to be done in an FEM model.

Since indentation is normally carried out to a relatively shallow depth, and in a region of limited lateral extent, any variations in the residual stress state within the region being tested can be neglected. Of course, there is scope for mapping changes in residual stress over the surface of a sample by making a number of indents in different locations. Furthermore, since there can be no stress normal to a free surface, the residual stress state in an indentation-tested region is fully characterized by the two (in-plane) principal stresses (see §2.2.2). In some cases, these two principle stresses will be the same, so that only a single value needs to be obtained. For example, stresses arising in a coating via differential thermal contraction between it and the substrate will often be isotropic in this way. Otherwise, three parameters are needed, two being values of the two principal stresses and the third specifying their orientation.

There have been many studies [60–63] aimed at measuring residual stresses via indentation, mostly based on the concept that, if the stress–strain relationship of the material is known, then it should be possible to evaluate the residual stress(es) from the indentation (plasticity) response. Many such studies have been carried out in the context of the effect of residual stresses on hardness, as described in §7.3.2. The ill-defined nature of hardness means that such approaches do not really have any potential in terms of experimental measurement of residual stresses. However, there have been a number of suggested approaches to the problem that do not involve hardness. Many of them [63–69] have been purely theoretical, but a few investigations have involved experimental indentation of samples in which controlled "artificial" residual stresses have been created by the external application of (equal or unequal biaxial) forces [69–72] or via differential thermal contraction between a substrate and a surface layer [73].

It can certainly be argued that, for such a procedure to be reliable, the way in which the indent is made, and in which the indentation outcome is processed, must be appropriate for obtaining the plasticity characteristics (in the absence of residual stress). As detailed elsewhere (§7.3.2, §8.2.1 and §8.2.2), many of the procedures proposed are subject to error (due to the deformed volume being too small to be representative of the bulk, the use of formulations that cannot adequately capture the stress and strain fields, use of "sharp" indenters or a focus on poorly defined parameters such as a hardness number). They are likely to be equally unreliable when extended to assessment of residual stresses.

However, there have been studies in which outcomes obtained using relatively large (spherical) indenters have been used to obtain insights into the influence of residual stresses on the indentation response. For example, using the experimental set-up shown in Fig. 8.15, and indenting two Al alloys (with a sphere of radius about 0.8 mm), while imposing a range of (unequal biaxial) residual stresses on the samples, Peng et al. [72] obtained the load–displacement plots shown in Fig. 8.16. It can be seen that these externally imposed ("residual") stresses had a significant influence on

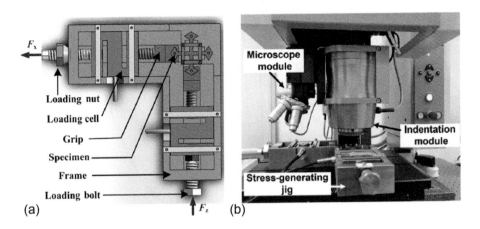

Fig. 8.15 Arrangement [72] for carrying out indentation while (unequal biaxial) stresses are being applied to samples, showing (a) a schematic representation of the loading frame and (b) a photo of the complete set-up.

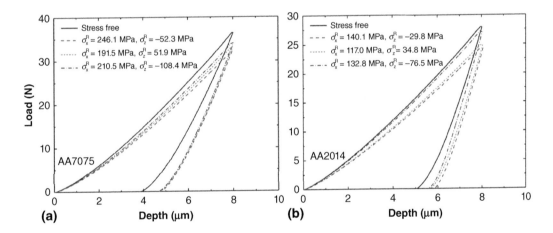

Fig. 8.16 Experimental load–displacement plots [72] obtained during indentation with a sphere of radius about 800 μm (to a depth of about 8 μm), with simultaneous application of the applied stresses shown, for Al alloy samples of: (a) 7075 ($\sigma_Y \sim$ 500 MPa) and (b) 2014 ($\sigma_Y \sim$ 300 MPa).

the experimental plots (despite the penetration ratio, δ/R, being only about 1%). Furthermore, the authors showed, using FEM modeling, that these outcomes were consistent with the imposed stress states. Of course, load–displacement plots cannot be used to infer anisotropic residual stress states, since they provide no directional information. The authors did, however, view the residual indents optically, noting that their shapes became elliptical when the residual stresses were unequal. On the other hand, they neither measured indent profiles nor used iterative FEM to infer residual stress states. Nevertheless, in terms of sensitivities, the plots in Fig. 8.16 offer some encouragement.

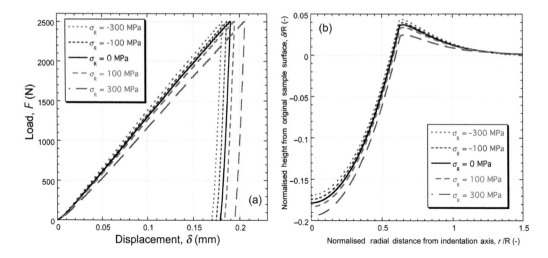

Fig. 8.17 FEM-predicted outcomes for indentation with a 1 mm radius sphere of an "artificial" alloy ($E = 200$ GPa, $\sigma_Y = 500$ MPa, $n = 0.5$ and $K = 500$ MPa) to a depth of about 200 μm, with equal biaxial residual stresses of 0 MPa, ±100 MPa and ±300 MPa, showing (a) load–displacement plots and (b) residual indent profiles.

In fact, it should be possible to infer residual stress levels from measured profiles, potentially extending to unequal biaxial states (if full 3-D profiles are extracted). Further information about the sensitivities involved is provided by Fig. 8.17, which shows FEM-predicted load–displacement plots and residual indent profiles for an "artificial" metal ($E = 200$ GPa, $\sigma_Y = 500$ MPa, $n = 0.5$ and $K = 500$ MPa), for indentation to a penetration ratio of about 20%, with five levels of (equal biaxial) residual stress (0, ±100 MPa and ±300 MPa). It can be seen that both are significantly affected, despite the fact that all of these levels are well below the yield stress.

It can be seen that tensile residual stresses promote penetration, whereas compressive stresses inhibit it (but to a lesser extent). This is expected from the nature of the yielding. Indenting initially creates mainly a compressive stress in the through-thickness direction: using either Tresca or von Mises criteria (§3.1.2), yielding will occur when the difference between this and the initial in-plane stress reaches the yield stress. Initial yielding will thus clearly be promoted by a tensile residual stress and inhibited by a compressive one. Of course, as penetration proceeds, the stress field becomes more complex, but the trends are still in the same direction and some asymmetry remains in terms of the magnitude, as well as the sense, of the effects with the two types of residual stress. There is further complexity with unequal biaxial residual stresses, but the main message of Fig. 8.17 (and Fig. 8.16) is that the sensitivities are such that inferring residual stresses in this way should be practicable. Of course, only profilometry has potential for establishing the directionality of any anisotropy in the residual stress state.

8.4.4 Indentation Creep Plastometry

A number of analytical or semi-analytical procedures [74–82] have been developed for obtaining creep characteristics from instrumented (depth-sensing) indentation data. Most commonly, the aim is to evaluate the stress exponent, n, that applies during the secondary creep regime. These procedures inevitably involve gross simplification of the complex stress and strain fields created as an indenter penetrates into a sample via creep deformation. In general, the outcomes (such as inferred n values) from such analytical procedures are very unreliable, for reasons that have become clear [83–86]. One of the reasons for this is that the primary regime of creep tends to influence the overall outcome throughout the test (because the creep strain field is continually expanding).

Iterative FEM simulation of the indentation process, on the other hand, offers the potential for accurate capture of evolving stress and strain fields, particularly if a constitutive law is used that covers both primary and secondary creep behavior. This is termed indentation creep plastometry. The procedure commonly involves the application of a constant load to a sample via a spherical indenter. The load is held at this level for an extended period, during which progressive penetration of the indenter into the sample is monitored. As with indentation plastometry, iterative FEM modeling of the process is carried out, with the creep characteristics (primary and secondary) captured in a constitutive law. Furthermore, the technique has the attraction of effectively investigating the creep response of the material over a range of stress levels during a single test (whereas separate tests are needed for different stress levels during conventional creep testing, which is described in §5.4.4). Also, the FEM simulation is based on true stress levels, whereas a drawback of conventional creep testing is that, unless the applied load is varied during the test (using a feedback loop based on the measured length change), the true stress changes progressively during the test as the specimen sectional area changes

A number of papers [87–93] have been published concerning this methodology. One of the main stumbling blocks during early work was the difficulty of carrying out an indentation creep test without stimulating (quasi-static) plasticity in the sample at the same time, particularly during the early stages of the test. However, this can be avoided by the prior production in the sample of a spherical cap recess with the same radius of curvature as the indenter [94]. Provided it has a suitable depth, this ensures that the (von Mises) stress in the sample doesn't reach the yield stress (for the temperature concerned) during the test.

An illustration of the methodology is provided by work [94] carried out on pure Ni samples at 750 °C (using a symmetrical arrangement involving a cermet sphere sandwiched between two identical samples). A tensile stress–strain curve for this material is shown in Fig. 8.18, where it can be seen that the yield stress at this temperature is about 67 MPa. Using the FEM mesh shown in Fig. 8.19(a), the predicted (elastic) von Mises stress field is shown in Fig. 8.19(b) for a load of 1 kN applied to a spherical indenter of radius 2 mm, with a recess depth of 1 mm. It can be seen that the yield stress is not reached anywhere within the sample.

8 Indentation Plastometry

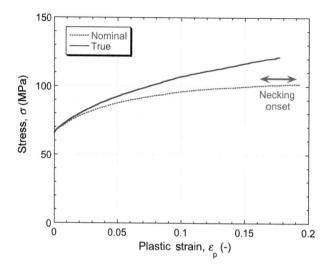

Fig. 8.18 Tensile stress–strain curves [94] for Ni at 750 °C, plotted as both nominal stress v. nominal strain and true stress v. true strain (obtained using Eqns. (3.10) and (3.11)), assuming that the stress and strain fields remained uniform throughout).

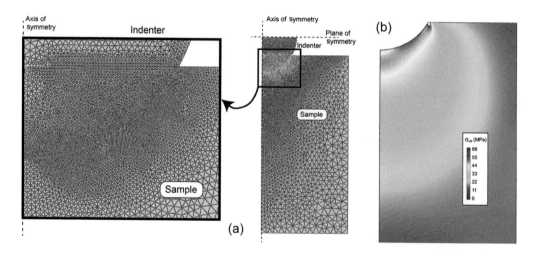

Fig. 8.19 (a) FEM mesh [94] for creep indentation and (b) predicted von Mises stress field within the sample for a load of 1 kN, an indenter of radius 2 mm and a prior spherical recess, having the same radius and a depth of 1 mm.

There is an issue with creep plastometry concerning the way in which the constitutive law is implemented in the FEM model. As mentioned above, it is important to use a law that covers both primary and secondary creep. Some of the options available are described in §3.2.4. The Miller–Norton law (Eqn. (3.17)) is often regarded as the most suitable for use in this type of FEM modeling. For a fixed temperature, it involves three adjustable parameters, C, m and n, so the number of degrees of freedom

8.4 Range of Indentation Plastometry Usage

during convergence on the "solution" is the same as for plasticity. This law provides a family of curves giving the strain as a function of time, each for a fixed stress level. Of course, during indentation, the stress level within the sample varies with both location and time during the test. A decision therefore has to be made about how the law is used to determine the strain history of each individual volume element, and hence the displacement history of the indenter.

The Miller–Norton equation can be differentiated with respect to time, to give

$$\frac{d\varepsilon_{cr}}{dt} = C\sigma^n t^m \tag{8.4}$$

The time can thus be expressed in terms of both strain rate and strain:

$$t = \left[\frac{\frac{d\varepsilon_{cr}}{dt}}{C\sigma^n}\right]^{1/m} = \left[\frac{(1+m)\varepsilon_{cr}}{C\sigma^n}\right]^{1/(1+m)} \tag{8.5}$$

Eliminating t and rearranging allows the strain rate to be expressed as a function of the strain:

$$\frac{d\varepsilon_{cr}}{dt} = \{C\sigma^n\}^{1/(1+m)}[(1+m)\varepsilon_{cr}]^{m/(1+m)} \tag{8.6}$$

It is then assumed that the cumulative creep strain defines the "state" of (a volume element of) the material, with the instantaneous creep strain rate determined by the current stress (in the volume element concerned) and the prior strain: the creep strain rate can thus be expressed solely as a function of the creep strain. This is depicted in Fig. 8.20. During each time increment, the net displacement of the indenter is found (within the FEM model) by monitoring the cumulative creep strain in each element up to that point, taking account of the (von Mises) stress in it, using Eqn. (8.6) to obtain the further strain that will arise in it during the time interval and then using compatibility conditions to solve and give the overall shape change of the domain. Quantification of the misfit between modeled and measured outcomes (displacement–time plots) and the convergence algorithm for finding the solution (best-fit set of parameter values in the Miller–Norton law) are analogous to those for the plasticity work.

The key objective is to obtain conventional (tensile) creep data (for any selected level of applied stress), at least in primary and secondary regimes, solely from indentation experiments – in fact, essentially from a single indentation experiment. The best-fit set of Miller–Norton parameter values can be used to predict the outcome of creep testing with any configuration, including, of course, the simple one of uniaxial tensile testing. In fact, for that case, it's not even necessary to carry out any further FEM modeling, since a tensile creep test is one in which the stress and strain field tends to remain homogeneous. (This is not true for compressive creep, when friction and barreling tend be significant.) A comparison between the outcome of a tensile creep experiment and a prediction based on indentation-derived values of the Miller–Norton parameters can therefore be made via simple manipulation of the Miller–Norton equation.

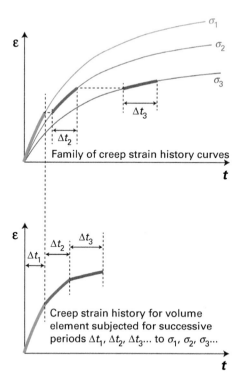

Fig. 8.20 Schematic illustration [94] of how the creep strain history of a volume element is assumed to be composed of a series of incremental strains, each dependent on the creep curve for the stress level concerned and the prior cumulative creep strain experienced by the element.

The outcome of such an operation can be seen in Fig. 8.21, where the experimental plots of the nominal creep strain as a function of time are compared with corresponding predicted plots obtained using the indentation-derived Miller–Norton parameter values. These predicted plots are based on the strain rate form of the Miller–Norton expression – i.e. Eqn. (8.6). This has been implemented by stepping through a series of time increments, calculating the latest strain rate by taking into account the changing value of the true stress. This is how the Miller–Norton expression should be used, since both the stress and the strain in it are true values.

The most striking feature of Fig. 8.21 is that the agreement between conventional tensile creep testing and the indentation-derived outcome is in general very good, at least within the primary and secondary regimes. The "tertiary" regime, which is quite noticeable with the highest level of applied stress, is not captured, even by using the Miller–Norton formulation in a way that takes account of the increasing level of true stress in such tests. The probable explanation for this discrepancy is that the true stress is starting to approach the yield stress. If this happens, then it is expected that the behavior will not be captured well using a creep model of this type, and plasticity characteristics (including the work hardening rate) are likely to have an effect. In fact,

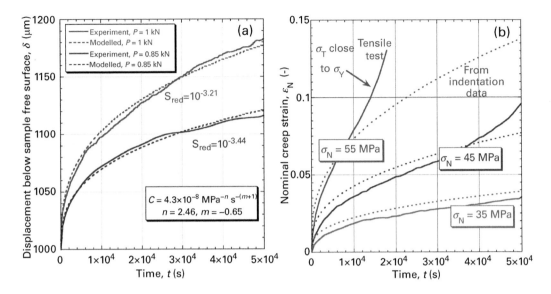

Fig. 8.21 Comparisons [94] between: (a) measured and (best-fit) modeled indenter penetration histories, with two applied loads (and showing the best-fit set of Miller–Norton parameter values and the values of the misfit parameter), and (b) creep strain curves from tensile testing, with fixed nominal stress, and those obtained using the best-fit Miller–Norton parameters and Eqn. (8.4).

any analytical formulation, such as the Miller–Norton law, is likely to be reliable only within a certain range of (true) stress.

It is of interest to note the range of stress and strain generated within an indentation test of this type, since it is clear that creep characteristics well outside of these ranges are unlikely to be captured well by such a test. Figure 8.22 shows fields of (von Mises) stress and creep strain within the sample at the end of the simulation with an applied nominal stress of 55 MPa. (This is actually after a time of 5×10^4 s, whereas the corresponding tensile creep test was stopped after about 1.8×10^4 s, when the strain rate was becoming very high.) This case therefore reflects a relatively severe test, in terms of generating high stresses and strains. It can immediately be seen, on comparing Fig. 8.22(a) with Fig. 8.19(b), that the stress levels have relaxed somewhat as the indenter has penetrated, and all of these stresses are well below the yield stress. It can also be seen from Fig. 8.22(b) that the creep strains generated within the sample range up to about 10–15%, which is appropriate for the comparisons shown in Fig. 8.21.

Finally, it can be seen in Fig. 8.22 that there is some "pile-up" around the indent, although it is not very pronounced. Of course, during conventional plastic deformation, such pile-ups can be quite noticeable, particularly for materials that exhibit little work hardening (allowing large plastic strains to develop near the pile-up). In general, while there is no clear analogue during creep deformation to a "work hardening" effect, there is a tendency for the stress and strain fields to become more "diffused" than during plastic deformation, such that pile-up (or "sink-in") effects are likely to be small.

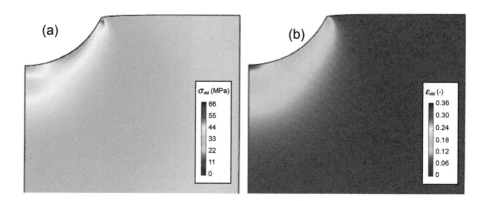

Fig. 8.22 Predicted [94] von Mises (a) stress and (b) strain fields within the sample 5×10^4 s after application of a load of 1 kN to an indenter of radius 2 mm, with a prior spherical recess having the same radius and a depth of 1 mm.

8.4.5 Indentation Superelastic Plastometry

The underlying methodology of indentation plastometry (i.e. iterative FEM simulation of the process, converging on a best-fit set of parameters in a constitutive law of some sort) can be applied to a range of mechanical response characteristics, although the main ones are those of plasticity, the presence of residual stresses and creep. It is probably safest to say that fracture characteristics cannot be obtained in this way, since crack propagation is highly constrained during indentation and is very difficult to capture reliably in an analytical relationship. However, one type of mechanical behavior that can in principle be tackled in this way is that of superelastic (SE) deformation, and the closely related shape memory effects. Constitutive laws can be formulated for SE deformation, as outlined in §3.2.3. The mechanism responsible for this type of behavior is that of strain arising from martensitic phase transformations, as described in §4.4.2.

Such deformation is considerably more complex than conventional plasticity (arising from dislocation motion and/or deformation twinning), as can be seen by comparing Fig. 3.7(b) with a typical plasticity stress–strain curve, such as those in Fig. 3.6. Moreover, it's not entirely clear whether the deformation should be regarded as elastic or plastic: strain levels are certainly much higher than those expected (for metals) in the elastic regime, but on the other hand they are at least largely reversible. Also, a constitutive law capturing the complete stress–strain curve, including the unloading response, must involve more parameters than those for conventional plasticity, as described in §3.2.3.

Nevertheless, in principle the methodology of indentation plasticity can be applied to SE deformation and indeed several publications describe attempts to do this, or at least to use FEM to simulate the indentation of SE alloys [95–98]. In most cases, commercial FEM packages were used and they do often incorporate standard SE formulations. This can be problematic for indentation studies of SE, since that formulation does not capture well the unloading response of regions subjected to a range of different strains, which is unavoidable during indentation. Moreover, there is a further point to note in this context, which is that the phase transformations concerned may involve a volume change, in which case there could be a dependence on the hydrostatic component of the stress state – see §3.1.1. This means that, strictly, the focus should not be solely on the von Mises stress and strain, as it normally is for conventional plasticity – see §6.3.1.

Some of these points are illustrated here using data from a study [98] based on a SE Ni-Ti alloy. In fact, information about the SE behavior of this alloy is provided in §3.2.3 – see Fig. 3.7, where a best-fit set of (seven) parameter values has been used to represent the loading and unloading curves (obtained in uniaxial compression). Predicted stress and strain fields during indentation, obtained using this set of values, for a load of 155 N (low enough to ensure that conventional plasticity was not stimulated), are shown in Fig. 8.23. This load stimulated complete transformation to martensitic in a region underneath the indenter, generating local strains of about 4%. The (von Mises) stress levels in this region are about 500–600 MPa. Relatively large regions of the sample have undergone only partial transformation to martensite, which is unavoidable during indentation.

Since SE indentation leaves little or no residual indent, the iterative FEM procedure can only be carried out using the load–displacement plot as target data. Moreover, reliable capture of the unloading part is not really possible using the standard formulation – see Fig. 3.7(b). While it is possible to carry out a convergence operation, the sensitivities are weak and the outcome unreliable. The upshot is that the indentation-derived stress–strain loop fits poorly to the (uniaxial) experimental plot during unloading. This is evident in Fig. 8.24.

Currently, therefore, use of indentation plastometry to obtain SE loops is not a reliable procedure, at least for the unloading part. Improvements will be dependent on new formulations to represent this part of the SE loop becoming available in commercial (or customized) FEM packages. Of course, it should be recognized that the driving force for such developments is relatively weak, since SE alloys are not in very widespread industrial use. Furthermore, the range of such alloys is rather limited and the need for characterization (via indentation) is far less than that for conventional metals. Nevertheless, the potential is there for characterization of SE behavior via indentation and a capability may emerge in due course.

8.4.6 Commercial Products

The concepts involved in indentation plastometry have been evolving for at least a couple of decades, but their routine implementation has been hampered by the need for iterative FEM simulation of the process. This presents multiple obstacles, since, not only must a model be set up (reflecting the experimental arrangement), and an algorithm developed for reliable and rapid convergence on the best-fit parameter set, but access to most commercial FEM packages carries (relatively high) financial overheads. Of course, these issues constitute strong drivers for turning to the

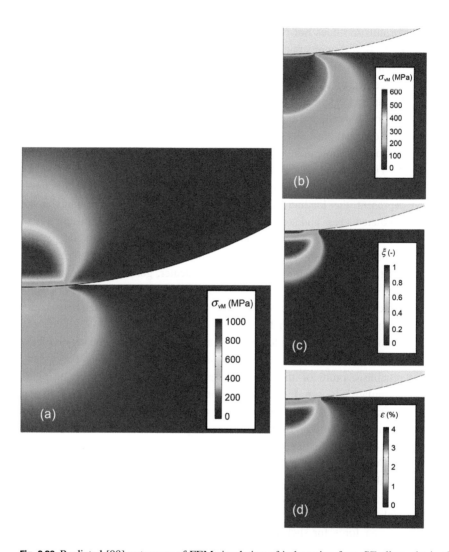

Fig. 8.23 Predicted [98] outcomes of FEM simulation of indentation for a SE alloy, obtained using the set of input parameters shown in Fig. 3.7(b), for an applied load of 155 N and a sphere radius of 2 mm. Fields are shown for: (a) the von Mises stress for both sample and spherical indenter, (b) the von Mises stress for sample (only), (c) the local volume fraction transformed to martensite and (d) the total (von Mises) strain.

alternative approach of some kind of analytical procedure for obtaining the stress–strain relationship from the indentation data. However, it is now clear that this cannot lead to reliable outcomes across the range of plasticity characteristics. Furthermore, the iterative FEM approach can be combined with FEM simulation of a tensile test, revealing information that is otherwise unavailable, such as the critical fracture strain – see §5.3.3.

It's clear that what is required is an integrated product, comprising an arrangement for carrying out the indentation and measuring the residual indent profile (or the load–

8.4 Range of Indentation Plastometry Usage

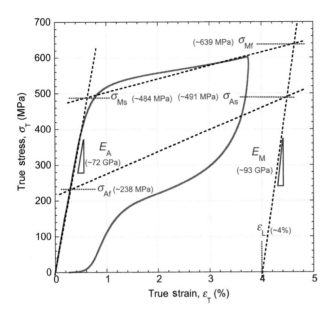

Fig. 8.24 Comparison [98] between the experimental stress–strain loop for a SE alloy and the SE parameter set inferred from an indentation load–displacement plot via a Nelder–Mead convergence operation.

displacement plot), combined with a software package that processes these experimental data to give final outcomes, such as a nominal stress–strain curve from a tensile test. It's important that the package should be stand-alone and user-friendly, requiring no knowledge of FEM (or access to any commercial FEM package). In principle, such a package could be supplied on its own, with the user carrying out the experimental measurements independently. However, in practice, this is not really viable, partly because none of the common types of loading system (nanoindenters, hardness testers and conventional testing frames) are ideally suited to the required range of load (few tens of N to a few kN), so suitable test set-ups are not readily accessible. Also, profilometers are rather specialized devices and they are not normally available as part of a loading set-up.

In fact, integrated facilities of this type are starting to become available. To take just one example, Plastometrex (https://plastometrex.com/) now offers such a product, with variants of several types in prospect. These include set-ups oriented towards study of creep, high strain rate plasticity, anisotropy and residual stresses. Figure 8.25(a) is a photo of an indentation plastometer in use, while Fig. 8.25(b) shows screenshots taken during operation of the software package. These show outcomes of FEM simulations of an indentation test and of a tensile test run using plasticity parameters values obtained from the indentation test. It's likely that facilities of this type will become widely used in the near future (and available in a similar cost range to that for hardness testers and conventional tensile machines – i.e. considerably cheaper than most "nanoindentation" facilities).

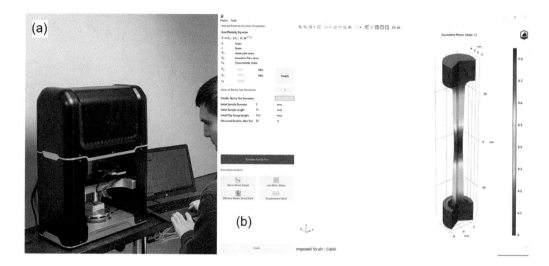

Fig. 8.25 An example of a commercial indentation plastometer, showing (a) the machine in use and (b) screenshots from the associated software package.

Appendix 8.1 Nelder–Mead Convergence Algorithm

Once an initial simplex has been created, each iteration comprises the following steps.

1. The values of S are calculated for each vertex and the vertices are ranked, such that $S(x_1) < S(x_2) < \cdots < S(x_{m+1})$. The point to be replaced is x_{m+1}. The centroid of the (reduced) simplex is calculated from

$$x_{cen} = \frac{1}{m}\sum_{j=1}^{m} x_j \qquad (8.7)$$

This defines the search direction $(x_{cen} - x_{m+1})$.

2. **Reflection**: A trial point is established by reflection of x_{m+1} through x_{cen}.

$$x_{ref} = x_{cen} + \alpha(x_{cen} - x_{m+1}) \qquad (8.8)$$

where α is a scale factor. The value of S is calculated for this point. If $S(x_1) < S(x_{ref}) < S(x_m)$, so that x_{ref} is of intermediate quality, then x_{ref} is accepted, replacing x_{m+1}. Otherwise, the algorithm proceeds to step 3.

3. **Expansion**: If $S(x_{ref}) < S(x_1)$, so that x_{ref} is the best point yet, this could indicate that the simplex is on an extended downward gradient and an expanded point is trialed:

$$x_{exp} = x_{cen} + \beta(x_{cen} - x_{m+1}) \qquad (8.9)$$

where β is a scale factor ($> \alpha$). The value of S is calculated for this point. If $S(x_{exp}) < S(x_{ref})$, then x_{exp} is accepted, replacing x_{m+1}. Otherwise, x_{ref} is accepted, replacing x_{m+1}.

4. **Outside contraction**: If $S(x_m) \leq S(x_{ref}) < S(x_{m+1})$, so that x_{ref} is an improvement on x_{m+1}, but would become the new worst point, the value of S is calculated for a point between x_{ref} and x_{cen}, called the outside contraction point:

$$x_{OC} = x_{cen} + \gamma(x_{cen} - x_{m+1}) \qquad (8.10)$$

where γ is a scale factor ($< \alpha$). The value of S is calculated for this point. If $S(x_{OC}) \leq S(x_{ref})$, then x_{OC} is accepted, replacing x_{m+1}. Otherwise, the algorithm proceeds to step 6.

5. **Inside contraction**: If $S(x_{m+1}) \leq S(x_{ref})$, so that x_{ref} is worse than all of the points in the existing simplex, then the value of S is calculated for a point between x_{cen} and x_{m+1}, called the inside contraction point:

$$x_{IC} = x_{cen} - \delta(x_{cen} - x_{m+1}) \tag{8.11}$$

where δ is another scale factor. The value of S is calculated for this point. If $S(x_{IC}) < S(x_{ref})$, then x_{IC} is accepted, replacing x_{m+1}. Otherwise, the algorithm proceeds to step 6.

6. **Shrink**: If none of the previous steps are able to improve the simplex, then it is shrunk towards the best vertex. This operation is defined by

$$x'_j = x_j + \delta(x_1 - x_j) \tag{8.12}$$

for $2 \leq j \leq (m+1)$. The algorithm then starts the next iteration at step 1.

The scale factors (α, β, γ and δ) are often ascribed values of 1, 2, 0.5 and 0.5, respectively, but these can be tuned to cope with particular situations, such as different levels of noise. The Scientific Python implementation allows for these scale factors to be adapted as the algorithm proceeds, as described in Gao and Han [52].

Appendix 8.2 Distribution of Plastic Work in terms of Strain Range

Within the FEM model, after each increment of strain, for each volume element, the stress, incremental strain and prior strain are recorded. The work done during that increment is evaluated (= stress × strain × volume) and that increment of work is associated with the strain concerned. Expressed mathematically, the increment of work done in the jth volume element during the kth increment of strain can be written

$$\Delta W_{j,k} = \sigma_{j,k} \Delta \varepsilon_{j,k} V_j \tag{8.13}$$

Clearly, the work done during the kth strain increment is given by

$$\Delta W_k = \sum_{j=1}^{M} \Delta W_{j,k} \tag{8.14}$$

where the summation is over the total number (M) of volume elements, and the total work done is

$$W_{\text{tot}} = \sum_{k=1}^{T} \Delta W_k \tag{8.15}$$

with this summation being over the total number (T) of strain increments. The total strain range is divided into a number of sub-ranges (bins) and the work done within each bin is then evaluated after a binning operation. This can be expressed as

$$W_{\text{bin},p} = \sum_{k=1}^{T} \sum_{j=1}^{M} \left(\Delta W_{j,k} f_{j,k,p} \right) \tag{8.16}$$

where $f_{j,k,p}$ is a function ascribed a value of 1 or 0, depending on whether the strain associated with the increment of work $\Delta W_{j,k}$ does or does not fall within the range of the pth bin.

References

1. Pelletier, H, Predictive model to estimate the stress–strain curves of bulk metals using nanoindentation. *Tribology International*, 2006. **39**: 593–606.
2. Heinrich, C, AM Waas and AS Wineman, Determination of material properties using nanoindentation and multiple indenter tips. *International Journal of Solids and Structures*, 2009. **46**: 364–376.
3. Dean, J, JM Wheeler and TW Clyne, Use of quasi-static nanoindentation data to obtain stress–strain characteristics for metallic materials. *Acta Materialia*, 2010. **58**: 3613–3623.
4. Guelorget, B, M Francois, C Liu and J Lu, Extracting the plastic properties of metal materials from microindentation tests: experimental comparison of recently published methods. *Journal of Materials Research*, 2007. **22**: 1512–1519.
5. Dao, M, N Chollacoop, KJ Van Vliet, TA Venkatesh and S Suresh, Computational modeling of the forward and reverse problems in instrumented sharp indentation. *Acta Materialia*, 2001. **49**: 3899–3918.
6. Bouzakis, K and N Michailidis, Coating elastic–plastic properties determined by means of nanoindentations and FEM-supported evaluation algorithms. *Thin Solid Films*, 2004. **469**: 227–232.
7. Bouzakis, K and N Michailidis, An accurate and fast approach for determining materials stress–strain curves by nanoindentation and its FEM-based simulation. *Materials Characterisation*, 2006. **56**: 147–157.
8. Bolzon, G, G Maier and M Panico, Material model calibration by indentation, imprint mapping and inverse analysis. *International Journal of Solids and Structures*, 2004. **41**(11–12): 2957–2975.
9. Bobzin, K, N Bagcivan, S Theiss, R Brugnara and J Perne, Approach to determine stress strain curves by FEM supported nanoindentation. *Materialwissenschaft Und Werkstofftechnik*, 2013. **44**(6): 571–576.
10. Patel, DK and SR Kalidindi, Correlation of spherical nanoindentation stress–strain curves to simple compression stress–strain curves for elastic–plastic isotropic materials using finite element models. *Acta Materialia*, 2016. **112**: 295–302.
11. Dean, J and TW Clyne, Extraction of plasticity parameters from a single test using a spherical indenter and FEM modelling. *Mechanics of Materials*, 2017. **105**: 112–122.
12. Campbell, JE, RP Thompson, J Dean and TW Clyne, Experimental and computational issues for automated extraction of plasticity parameters from spherical indentation. *Mechanics of Materials*, 2018. **124**: 118–131.
13. Campbell, JE, RP Thompson, J Dean and TW Clyne, Comparison between stress–strain plots obtained from indentation plastometry, based on residual indent profiles, and from uniaxial testing. *Acta Materialia*, 2019. **168**: 87–99.
14. Taljat, B, T Zacharia and F Kosel, New analytical procedure to determine stress–strain curve from spherical indentation data. *International Journal of Solids and Structures*, 1998. **35**(33): 4411–4426.
15. Herbert, EG, GM Pharr, WC Oliver, BN Lucas and JL Hay, On the measurement of stress–strain curves by spherical indentation. *Thin Solid Films*, 2001. **398**: 331–335.
16. Pelletier, H, Predictive model to estimate the stress–strain curves of bulk metals using nanoindentation. *Tribology International*, 2006. **39**(7): 593–606.
17. Kang, BSJ, Z Yao and EJ Barbero, Post-yielding stress–strain determination using spherical indentation. *Mechanics of Advanced Materials and Structures*, 2006. **13**(2): 129–138.

18. Basu, SM, A Moseson and MW Barsoum, On the determination of spherical nanoindentation stress–strain curves. *Journal of Materials Research*, 2006. **21**(10): 2628–2637.
19. Hausild, P, A Materna and J Nohava, On the identification of stress–strain relation by instrumented indentation with spherical indenter. *Materials & Design*, 2012. **37**: 373–378.
20. Hamada, AS, FM Haggag and DA Porter, Non-destructive determination of the yield strength and flow properties of high-manganese twinning-induced plasticity steel. *Materials Science and Engineering A: Structural Materials Properties Microstructure and Processing*, 2012. **558**: 766–770.
21. Xu, BX and X Chen, Determining engineering stress–strain curve directly from the load–depth curve of spherical indentation test. *Journal of Materials Research*, 2010. **25**(12): 2297–2307.
22. Pathak, S and SR Kalidindi, Spherical nanoindentation stress–strain curves. *Materials Science & Engineering R: Reports*, 2015. **91**: 1–36.
23. Pintaude, G and AR Hoechele, Experimental analysis of indentation morphologies after spherical indentation. *Materials Research*, 2014. **17**: 56–60.
24. Futakawa, M, T Wakui, Y Tanabe and I Ioka, Identification of the constitutive equation by the indentation technique using plural indenters with different apex angles. *Journal of Materials Research*, 2001. **16**: 2283–2292.
25. Chollacoop, N, M Dao and S Suresh, Depth-sensing instrumented indentation with dual sharp indenters. *Acta Materialia*, 2003. **51**(13): 3713–3729.
26. Cheng, Y-T and C-M Cheng, Scaling, dimensional analysis, and indentation measurements. *Materials Science and Engineering: R: Reports*, 2004. **44**(4–5): 91–149.
27. Capehart, TW and YT Cheng, Determining constitutive models from conical indentation: sensitivity analysis. *Journal of Materials Research*, 2003. **18**(4): 827–832.
28. Bucaille, JL, S Stauss, E Felder and J Michler, Determination of plastic properties of metals by instrumented indentation using different sharp indenters. *Acta Materialia*, 2003. **51**(6): 1663–1678.
29. Ma, ZS, YC Zhou, SG Long, XL Zhong and C Lu, Characterization of stress–strain relationships of elastoplastic materials: an improved method with conical and pyramidal indenters. *Mechanics of Materials*, 2012. **54**: 113–123.
30. Sun, Y, S Zheng, T Bell and J Smith, Indenter tip radius and load frame compliance calibration using nanoindentation loading curves. *Philosophical Magazine Letters*, 1999. **79**(9): 649–658.
31. Ullner, C, E Reimann, H Kohlhoff and A Subaric-Leitis, Effect and measurement of the machine compliance in the macro range of instrumented indentation test. *Measurement*, 2010. **43**(2): 216–222.
32. Van Vliet, KJ, L Prchlik and JF Smith, Direct measurement of indentation frame compliance. *Journal of Materials Research*, 2011. **19**(1): 325–331.
33. Campbell, JE, T Kalfhaus, R Vassen, RP Thompson, J Dean and TW Clyne, Mechanical properties of sprayed overlayers on superalloy substrates, obtained via indentation testing. *Acta Materialia*, 2018. **154**: 237–245.
34. Lee, J, C Lee and B Kim, Reverse analysis of nano-indentation using different representative strains and residual indentation profiles. *Materials & Design*, 2009. **30**(9): 3395–3404.
35. Yao, WZ, CE Krill, B Albinski, HC Schneider, and JH You, Plastic material parameters and plastic anisotropy of tungsten single crystal: a spherical micro-indentation study. *Journal of Materials Science*, 2014. **49**(10): 3705–3715.

36. Wang, MZ, JJ Wu, Y Hui, ZK Zhang, XP Zhan and RC Guo, Identification of elastic–plastic properties of metal materials by using the residual imprint of spherical indentation. *Materials Science and Engineering A: Structural Materials Properties Microstructure and Processing*, 2017. **679**: 143–154.
37. Richmond, JC and AC Francisco, Use of plastic replicas in evaluating surface texture of enamels. *Journal of Research of the National Bureau of Standards*, 1949. **42**(5): 449–460.
38. Skelton, RP, HJ Maier and HJ Christ, The Bauschinger effect, Masing model and the Ramberg–Osgood relation for cyclic deformation in metals. *Materials Science and Engineering: A Structural Materials: Properties, Microstructure and Processing*, 1997. **238**(2): 377–390.
39. Samuel, KG and P Rodriguez, On power-law type relationships and the Ludwigson explanation for the stress–strain behaviour of AISI 316 stainless steel. *Journal of Materials Science*, 2005. **40**(21): 5727–5731.
40. Belytschko, T, R Gracie and G Ventura, A review of extended/generalized finite element methods for material modeling. *Modelling and Simulation in Materials Science and Engineering*, 2009. **17**(4).
41. Roters, F, P Eisenlohr, L Hantcherli, DD Tjahjanto, TR Bieler and D Raabe, Overview of constitutive laws, kinematics, homogenization and multiscale methods in crystal plasticity finite-element modeling: theory, experiments, applications. *Acta Materialia*, 2010. **58**(4): 1152–1211.
42. Giannakopoulos, AE and S Suresh, Determination of elastoplastic properties by instrumented sharp indentation. *Scripta Materialia*, 1999. **40**(10): 1191–1198.
43. Taljat, B and GM Pharr, Development of pile-up during spherical indentation of elastic–plastic solids. *International Journal of Solids and Structures*, 2004. **41**(14): 3891–3904.
44. Karthik, V, P Visweswaran, A Bhushan, DN Pawaskar, KV Kasiviswanathan, T Jayakumar and B Raj, Finite element analysis of spherical indentation to study pile-up/sink-in phenomena in steels and experimental validation. *International Journal of Mechanical Sciences*, 2012. **54**(1): 74–83.
45. Klein, CA, Anisotropy of Young modulus and Poisson ratio in diamond. *Materials Research Bulletin*, 1992. **27**(12): 1407–1414.
46. Isselin, J, A Iost, J Golek, D Najjar, and M Bigerelle, Assessment of the constitutive law by inverse methodology: small punch test and hardness. *Journal of Nuclear Materials*, 2006. **352**(1–3): 97–106.
47. Swaddiwudhipong, S, J Hua, E Harsono, ZS Liu and NSB Ooi, Improved algorithm for material characterization by simulated indentation tests. *Modelling and Simulation in Materials Science and Engineering*, 2006. **14**(8): 1347–1362.
48. Peyrot, I, PO Bouchard, R Ghisleni and J Michler, Determination of plastic properties of metals by instrumented indentation using a stochastic optimization algorithm. *Journal of Materials Research*, 2009. **24**(3): 936–947.
49. Chen, J, HN Chen and J Chen, Evaluation of mechanical properties of structural materials by a spherical indentation based on the representative strain – an improved algorithm at great depth ratio. *Acta Metallurgica Sinica – English Letters*, 2011. **24**(5): 405–414.
50. Meng, L, P Breitkopf, B Raghavan, G Mauvoisin, O Bartier and X Hernot, On the study of mystical materials identified by indentation on power law and Voce hardening solids. *International Journal of Material Forming*, 2019. **12**: 587–602.
51. Nelder, JA and R Mead, A simplex method for function minimization. *The Computer Journal*, 1965. **7**(4): 308–313.

52. Gao, FC and LX Han, Implementing the Nelder–Mead simplex algorithm with adaptive parameters. *Computational Optimization and Applications*, 2012. **51**(1): 259–277.
53. Oliphant, TE, Python for scientific computing. *Computing in Science & Engineering*, 2007. **9**(3): 10–20.
54. van der Walt, S, SC Colbert and G Varoquaux, The Numpy array: a structure for efficient numerical computation. *Computing in Science & Engineering*, 2011. **13**(2): 22–30.
55. Bunge, HJ, *Texture Analysis in Materials Science: Mathematical Methods*. London: Butterworth, 1982.
56. Wenk, HR and P Van Houtte, Texture and anisotropy. *Reports on Progress in Physics*, 2004. **67**(8): 1367–1428.
57. Zhao, Z, W Mao, F Roters and D Raabe, A texture optimization study for minimum earing in aluminium by use of a texture component crystal plasticity finite element method. *Acta Materialia*, 2004. **52**(4): 1003–1012.
58. Raabe, D, Y Wang and F Roters, Crystal plasticity simulation study on the influence of texture on earing in steel. *Computational Materials Science*, 2005. **34**(3): 221–234.
59. Clyne, TW and PJ Withers, *An Introduction to Metal Matrix Composites*. Cambridge Solid State Science Series, Davis, E and I Ward, eds. Cambridge: Cambridge University Press, 1993.
60. Taljat, B and GM Pharr. Measurement of residual stresses by load and depth sensing spherical indentation, in *Thin Films: Stresses and Mechanical Properties VIII*. Warrendale, PA: Materials Research Society, 2000, pp. 519–524.
61. Swadener, JG, B Taljat and GM Pharr, Measurement of residual stress by load and depth sensing indentation with spherical indenters. *Journal of Materials Research*, 2001. **16**(7): 2091–2102.
62. Jang, JI, Estimation of residual stress by instrumented indentation: a review. *Journal of Ceramic Processing Research*, 2009. **10**(3): 391–400.
63. Sakharova, NA, PA Prates, MC Oliveira, JV Fernandes and JM Antunes, A simple method for estimation of residual stresses by depth-sensing indentation. *Strain*, 2012. **48**(1): 75–87.
64. Cao, YP and J Lu, A new method to extract the plastic properties of metal materials from an instrumented spherical indentation loading curve. *Acta Materialia*, 2004. **52**: 4023–4032.
65. Xu, ZH and XD Li, Influence of equi-biaxial residual stress on unloading behaviour of nanoindentation. *Acta Materialia*, 2005. **53**(7): 1913–1919.
66. Larsson, PL, On the influence of elastic deformation for residual stress determination by sharp indentation testing. *Journal of Materials Engineering and Performance*, 2017. **26**(8): 3854–3860.
67. Zhang, TH, C Yu, GJ Peng and YH Feng, Identification of the elastic–plastic constitutive model for measuring mechanical properties of metals by instrumented spherical indentation test. *MRS Communications*, 2017. **7**(2): 221–228.
68. Wang, ZY, LX Deng and JP Zhao, A novel method to extract the equi-biaxial residual stress and mechanical properties of metal materials by continuous spherical indentation test. *Materials Research Express*, 2019. **6**(3).
69. Zhang, TH, WQ Cheng, GJ Peng, Y Ma, WF Jiang, JJ Hu and H Chen, Numerical investigation of spherical indentation on elastic-power-law strain-hardening solids with non-equibiaxial residual stresses. *MRS Communications*, 2019. **9**(1): 360–369.
70. Pham, TH and SE Kim, Determination of equi-biaxial residual stress and plastic properties in structural steel using instrumented indentation. *Materials Science and Engineering: A Structural Materials: Properties, Microstructure and Processing*, 2017. **688**: 352–363.

71. Peng, GJ, ZK Lu, Y Ma, YH Feng, Y Huan and TH Zhang, Spherical indentation method for estimating equibiaxial residual stress and elastic–plastic properties of metals simultaneously. *Journal of Materials Research*, 2018. **33**(8): 884–897.
72. Peng, GJ, FG Xu, JF Chen, HD Wang, JJ Hu and TH Zhang, Evaluation of non-equibiaxial residual stresses in metallic materials via instrumented spherical indentation. *Metals*, 2020. **10**.
73. Dean, J, G Aldrich-Smith and TW Clyne, Use of nanoindentation to measure residual stresses in surface layers. *Acta Materialia*, 2011. **59**(7): 2749–2761.
74. Liu, H, Y Chen, Y Tang, S Wei and G Nuiu, Tensile and indentation creep behaviour of Mg-5%Sn and Mg-5%Sn-2%Di alloys. *Materials Science and Engineering A*, 2007. **464**: 124–128.
75. Takagi, H, M Dao and M Fujiwara, Analysis on pseudo-steady indentation creep. *Acta Mechanica Solida Sinica*, 2008. **21**: 283–288.
76. Marques, VMF, B Wunderle, C Johnston and PS Grant, Nanomechanical characterisation of Sn-Ag-Cu/Cu joints – part 2: nanoindentation creep and its relationship with uniaxial creep as a function of temperature. *Acta Materialia*, 2013. **61**(7): 2471–2480.
77. Geranmayeh, AR and R Mahmudi, Indentation creep of a cast Mg-6Al-1Zn-0.7Si alloy. *Materials Science and Engineering: A Structural Materials: Properties, Microstructure and Processing*, 2014. **614**: 311–318.
78. Chatterjee, A, M Srivastava, G Sharma and JK Chakravartty, Investigations on plastic flow and creep behaviour in nano and ultrafine grain Ni by nanoindentation. *Materials Letters*, 2014. **130**: 29–31.
79. Wang, Y and J Zeng, Effects of Mn addition on the microstructure and indentation creep behaviour of the hot dip Zn coating. *Materials & Design*, 2015. **69**: 64–69.
80. Mahmudi, R, M Shalbafi, M Karami and AR Geranmayeh, Effect of Li content on the indentation creep characteristics of cast Mg-Li-Zn alloys. *Materials & Design*, 2015. **75**: 184–190.
81. Ma, Y, GJ Peng, DH Wen and TH Zhang, Nanoindentation creep behavior in a CoCrFeCuNi high-entropy alloy film with two different structure states. *Materials Science and Engineering: A Structural Materials: Properties, Microstructure and Processing*, 2015. **621**: 111–117.
82. Ginder, RS, WD Nix and GM Pharr, A simple model for indentation creep. *Journal of the Mechanics and Physics of Solids*, 2018. **112**: 552–562.
83. Goodall, R and TW Clyne, A critical appraisal of the extraction of creep parameters from nanoindentation data obtained at room temperature. *Acta Materialia*, 2006. **54**(20): 5489–5499.
84. Chen, J and SJ Bull, The investigation of creep of electroplated Sn and Ni-Sn coating on copper at room temperature by nanoindentation. *Surface and Coatings Technoology*, 2009. **203**(12): 1609–1617.
85. Dean, J, J Campbell, G Aldrich-Smith and TW Clyne, A critical assessment of the "stable indenter velocity" method for obtaining the creep stress exponent from indentation data. *Acta Materialia*, 2014. **80**: 56–66.
86. Campbell, J, J Dean and TW Clyne, Limit case analysis of the "stable indenter velocity" method for obtaining creep stress exponents from constant load indentation tests. *Mechanics of Time-dependent Materials*, 2016. **1**: 31–43.
87. Liu, YJ, B Zhao, BX Xu and ZF Yue, Experimental and numerical study of the method to determine the creep parameters from the indentation creep testing. *Materials Science and*

Engineering: A Structural Materials: Properties, Microstructure and Processing, 2007. **456**(1–2): 103–108.
88. Galli, M and ML Oyen, Spherical indentation of a finite poroelastic coating. *Applied Physics Letters*, 2008. **93**(3).
89. Wu, JL, Y Pan and JH Pi, On indentation creep of two Cu-based bulk metallic glasses via nanoindentation. *Physica B: Condensed Matter*, 2013. **421**: 57–62.
90. Dean, J, A Bradbury, G Aldrich-Smith and TW Clyne, A procedure for extracting primary and secondary creep parameters from nanoindentation data. *Mechanics of Materials*, 2013. **65**: 124–134.
91. Su, CJ, EG Herbert, S Sohn, JA LaManna, WC Oliver and GM Pharr, Measurement of power-law creep parameters by instrumented indentation methods. *Journal of the Mechanics and Physics of Solids*, 2013. **61**(2): 517–536.
92. Cordova, ME and YL Shen, Indentation versus uniaxial power-law creep: a numerical assessment. *Journal of Materials Science*, 2015. **50**(3): 1394–1400.
93. Rickhey, F, JH Lee and H Lee, An efficient way of extracting creep properties from short-time spherical indentation tests. *Journal of Materials Research*, 2015. **30**(22): 3542–3552.
94. Burley, M, JE Campbell, J Dean and TW Clyne, A methodology for obtaining primary and secondary creep characteristics from indentation experiments, using a recess. *International Journal of Mechanical Sciences*, 2020. **176**: 105577
95. Muir Wood, AJ and TW Clyne, Measurement and modelling of the nanoindentation response of shape memory alloys. *Acta Materialia*, 2006. **54**(20): 5607–5615.
96. Neupane, R and Z Farhat, Prediction of indentation behavior of superelastic TiNi. *Metallurgical and Materials Transactions A: Physical Metallurgy and Materials Science*, 2014. **45A**(10): 4350–4360.
97. Frost, M, A Kruisova, V Shanel, P Sedlak, P Hausild, M Kabla, D Shilo and M Landa, Characterization of superelastic NiTi alloys by nanoindentation: experiments and simulations. *Acta Physica Polonica A*, 2015. **128**(4): 664–669.
98. Roberto-Pereira, FF, JE Campbell, J Dean and TW Clyne, Extraction of superelasticity parameter values from instrumented indentation via iterative FEM modelling. *Mechanics of Materials*, 2019. **134**: 143–152.

9 Nanoindentation and Micropillar Compression

Mechanical testing on a very fine scale, particularly indentation, has become extremely popular. Sophisticated equipment has been developed, often with accompanying software that facilitates the extraction of properties such as stiffness, hardness and other plasticity parameters. The region being tested can be very small – down to sub-micron dimensions. However, strong caveats should be noted concerning such measurements, particularly relating to plasticity. Some of these concern various potential sources of error, such as the effects of surface roughness, oxide films, uncertainty about the precise geometry of the indenter tip etc. Moreover, even if these can be largely eliminated, extraneous effects tend to arise when (plastically) deforming a small region that is constrained by surrounding (elastic) material. They are often grouped together under the heading of "size effects," with a clear tendency observed for material to appear harder as the scale of the testing is reduced. Various explanations for this have been put forward, some based on dislocation characteristics, but understanding is incomplete and compensating for them in a systematic way does not appear to be viable. A similar level of uncertainty surrounds the outcome of fine scale uniaxial compression testing, although the conditions, and the sources of error, are rather different from those during nanoindentation. Despite the attractions of these techniques, and the extensive work done with them, they are thus of limited use for the extraction of meaningful mechanical properties (related to plasticity).

9.1 General Background

Growth in the usage and general profile of *"**nanoindentation**"* has been remarkable over the past two or three decades. There is, however, an issue concerning definition of the term. Since the indentation technique dates back well over a century, and the idea of carrying it out on at least a relatively fine scale is also far from novel – in effect originating with the Vickers hardness test of 1924 (see §7.2.3) – it's not immediately clear what innovation prompted this explosion of interest. In practice, it is not really refinement of the scale of the indentation that has been pivotal, but rather development of the technology for continuously monitoring the (progressively increasing) load and, particularly, the resultant indenter penetration during the test. For this reason, the term *"**depth-sensing indentation**"* is sometimes preferred when describing the technique.

It offers the potential for obtaining information not accessible via simple measurement of the final depth or lateral dimensions of an indent. To illustrate the level of interest that this has stimulated, an Oliver and Pharr paper [1] of 1992, outlining how a sample stiffness value can be obtained from a load–displacement plot, had received over 16,000 citations by 2020 – a figure still increasing by over 1,000 pa: this is reputedly the most highly cited paper in the whole domain of materials science and engineering. The number of papers with the term "nanoindentation" in the title or abstract (which started to rise in the late 1980s) stood at about 27,000 in mid-2020, although the rate of publication of such papers has been fairly constant over the past decade or so and has probably passed its peak.

Of course, there has certainly been an increasing emphasis on carrying out this type of test on a very fine scale. This is often considered to offer potential for obtaining fundamental information, as well as allowing study of very small regions – for example, individual constituents in a microstructure or thin surface layers or coatings. In practice, however, such characterization is often severely hampered by various extraneous effects that can arise with very fine scale measurements.

The main issues involved in carrying out nanoindentation, and interpreting the data obtained, are covered in a number of review articles [2–8], and also in a few books [9] and edited multi-author volumes [10–12]. Many of these review papers focus quite strongly on the so-called "*size effect*" – i.e. an apparent dependence of the outcome (usually an inferred mechanical property of some sort, such as a yield stress) on the size of the indent – see §9.4. It's also worth noting that there has been parallel development [13–15] of the "*micropillar compression test*." This involves a very different testing configuration, and brings different types of challenge in sample preparation, data interpretation etc., but there is commonality in the sense that it also is dependent on accurate measurement of small loads and displacements. In practice, the same experimental facilities are often used for both techniques. Furthermore, some of the same issues, such as a size effect, arise in both.

It is perhaps worth noting that both nanoindentation and micropillar compression are often applied to non-metallic materials. In many such cases, plasticity takes place less readily than in metals and the relative importance of factors such as **prior dislocation density**, **dislocation mobility**, the presence of **oxide layers**, **strain rate**, **fracture toughness**, **anisotropy** etc. may differ from that typical of metallic systems. Since the focus of this book is on metals (particularly their plasticity characteristics), limited attention is paid to such cases here. However, an important point about both of these techniques is that the region being mechanically interrogated is commonly (although not exclusively) a **single crystal**. Since their plastic deformation is quite different from that of corresponding polycrystals (see §4.3.1), with **crystallographic orientation** being a key factor, anisotropy is often inherent in the sample response (for both metallic and non-metallic materials).

9.2 Nanoindentation Equipment

9.2.1 Equipment Design

The development of nanoindentation has been largely dependent on a capability for good control over the applied load and accurate measurement of the displacement of the indenter tip during penetration. It's not entirely clear what the prefix "nano-" actually means in this context, but the displacement resolution certainly needs to be in the nm range (and in some cases a sub-nm resolution is claimed). Of course, even highly polished samples often have a surface roughness of at least several tens of nm, so the significance of a resolution in that range is doubtful in terms of absolute depth values. The minimum penetration likely to be created is of the order of a few tens of nm and a more typical value, even for fine scale work, is probably around 100 nm. The corresponding load will depend on indenter shape and sample hardness, but would usually be of the order of a few mN – certainly no less than several hundred μN. Of course, even 1 mN is a very low load and a resolution of 1 nm is certainly demanding, so precision equipment is needed. On the other hand, there are sometimes requirements for much higher loads than this and some "nanoindenter" loading systems (usually solenoid-based – see below) can generate forces of up to, say, 20 N. Even such values, however, are well below those typically needed for indentation plastometry (Chapter 8), which are normally created using a leadscrew and crosshead arrangement similar to those commonly used in conventional tensile testing.

A schematic representation of a typical loading and measurement configuration [16] is shown in Fig. 9.1. As indicated in the figure, the force is commonly applied and controlled via the current in a solenoid (located in the field of a permanent magnet). This gives good linearity between current and force, although, at least for relatively large forces, there is often a *heating effect* (and *thermal drift* is potentially a major problem area for nanoindenters – see §9.2.3). An alternative is to apply electrostatic force [17] via the field between two electrodes acting on a central plate attached to the indenter, with a voltage being applied between it and one of the electrodes. This is more compact and generates less heat than an electromagnetic arrangement, but the

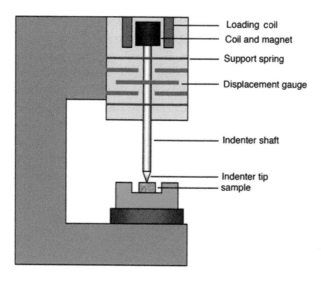

Fig. 9.1 Schematic representation [16] of the loading and measurement configuration of a typical nanoindenter.

force is very limited (~0.1 mN). A third possibility is a spring-based system, although the resultant force tends to be non-linear with distance and sensitive to temperature, misalignment etc.

The displacement is often measured from the ***capacitance*** of the region between a plate attached to the indenter shaft and another that is attached to the loading frame. This gives good resolution (~1 nm), albeit over a relatively limited range (perhaps just a few hundred nm). Also, it may be necessary to take account of the compliance of the loading train when determining the actual penetration of the indenter into the sample from the displacement reading. This is, of course, a similar problem to that encountered when applying compliance calibration during conventional tensile and compressive testing (§5.2.2). The relative significance of the machine compliance may be lower for nanoindentation, but the resolution being sought is often much higher than for tensile testing, so this correction may still be necessary. As would be expected, there are various trade-offs and compromises involved in the design and operation of a nanoindenter, details of which are available in the literature [5, 16, 18–21].

9.2.2 Nanoindenter Tips

Most nanoindenter tips are made of ***diamond*** or a form of ***alumina*** (such as ***sapphire***). There are in fact a number of alternatives, which might in particular be relevant if usage is being considered at elevated temperatures (where diamond can oxidize rapidly in air) or in other aggressive environments. Figure 9.2 shows the hardness and stiffness of several candidate tip materials, as a function of temperature [22]. These data do highlight the major advantage offered by diamond in terms of being harder and stiffer (over the complete temperature range), although the toughness – a more complex property that is often difficult to measure reliably – may also be relevant.

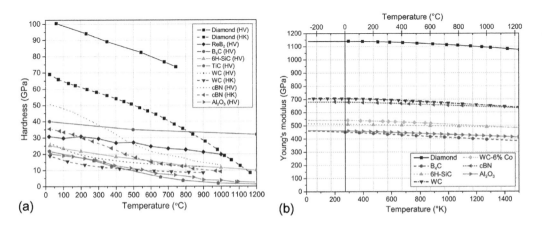

Fig. 9.2 Measured values [22] of (a) hardness (Vickers and Knoop) and (b) Young's modulus of candidate indenter materials, as a function of temperature.

Nevertheless, in general high values of E and H are beneficial, particularly since nanoindentation is commonly applied to hard materials and coatings. If diamond is ruled out on grounds of susceptibility to oxidation or other chemical attack (such as forming a carbide with a metal sample), then alumina is usually preferred. It's clear from Fig. 9.2 that there are several alternative materials with higher stiffness and hardness, but alumina is usually more chemically inert and stable than all of them and so is often used when this is a key requirement. However, as outlined in Chapter 8, the situation is different with indentation plastometry, for which large spherical indenters are used. In that case, a *cermet* (commonly WC with a Co binder) is usually preferred, since they offer the most attractive combination of mechanical properties, cost and ease of manufacture.

Of course, the other main issue relating to choice of indenter concerns the geometry and scale of the tip. Most tips used in nanoindenters are of the "sharp" type, a term that usually implies the existence of flat surfaces, edges and an apex. There are two main motivations for this. One is that it allows significant penetration, even into hard samples, while the applied force remains relatively low (and, as mentioned above, the creation of large forces in a high precision set-up is problematic). The other is that they are relatively easy to manufacture. Machining of such hard materials naturally presents challenges, but it is easier to cut a small number of well-defined faces than to create the main alternatives of a spherical or cono-spheroidal shape, particularly if they have a small radius. In fact, as mentioned in §7.2.3, the main attraction of using a Berkovich (three-sided pyramid), rather than a Vickers (four-sided pyramid), is simply that it is easier to machine (so as to create a well-defined apex).

However, there are clear downsides to use of "sharp" indenters. One, which was also mentioned in §7.2.3 (Fig. 7.11), is that they are often prone to becoming damaged – particularly along the edges and at the apex. This can arise from mechanical wear/micro-fracture or, particularly in the case of diamond, from local *oxidation* or other chemical reaction (at elevated temperature). The fine scale of the tip (i.e. the fact that penetration into the sample is often very shallow) means that even a small degree of such damage can dramatically alter the nanoindentation (load–displacement) response. An example of such damage can be seen in Fig. 9.3, which shows [23] the effect of a single indentation operation on a Berkovich tip. While the "blunting" effect appears to be relatively minor, the change in apex shape shown at the bottom of this figure is such that, for subsequent fine scale measurements of various types, a significant error would be introduced. Nanoindenter tips should be inspected frequently (via SEM and/or AFM) for such damage, although this is not always done.

There are also issues relating to FEM modeling of the nanoindentation process. As highlighted in §8.3 (in the context of indentation plastometry), this is a potentially powerful approach to the extraction of information from indentation experiments. However, with "sharp" indenters and fine scale indentation, certain difficulties commonly arise. Physical damage to the tip introduces errors into correlation between experiment and model. Moreover, the FEM simulation itself can present problems in terms of specification of the mesh domain and boundary conditions in the vicinity of edges and apices. There is also an increase in complexity associated with the

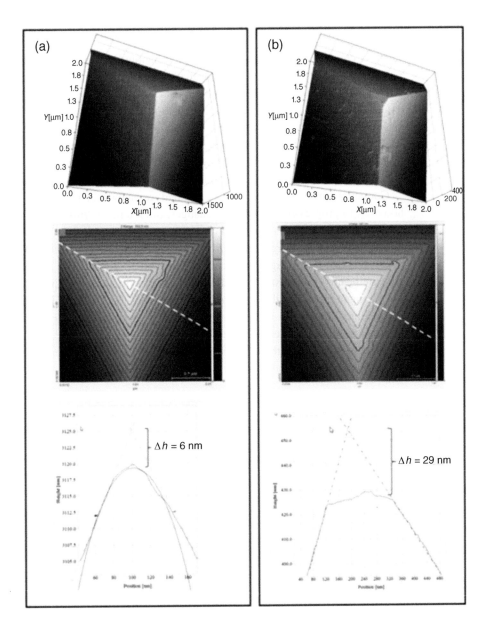

Fig. 9.3 AFM data [23] taken from a diamond Berkovich tip (a) in new condition and (b) after a single indentation into a SiC sample at 200 °C.

requirement for 3-D simulation (in the absence of radial symmetry). Finally, the potential significance of surface roughness, oxide films etc. is much greater with fine scale indentation and such effects are very difficult to incorporate into FEM models. For these reasons, while FEM simulation of nanoindentation has frequently been undertaken, there is often a large degree of uncertainty surrounding the significance of the outcomes.

9.2.3 High Temperature Testing and Controlled Atmosphere Operation

There is considerable interest in carrying out nanoindentation at high temperatures. This is to some extent facilitated by the small scale of both sample and loading system, but the high resolution required for the displacement data means that ***thermal drift*** (and the associated dimensional changes due to thermal expansion) can be a serious problem [24, 25]. Even in a relatively stable set-up operating at ambient temperature, there are commonly progressive changes in measured displacement, typically at rates of the order of 1 nm s^{-1}. Since, for many purposes, nm-scale resolution is required, errors arising from this source over quite short periods can be problematic.

Correction procedures [25] can be applied, as shown in Fig. 9.4. These relate, however, to progressive background changes, often termed "***frame drift***." Correcting for dimensional changes that arise from heat flow stimulated by contact between sample and indenter ("***contact drift***") is more difficult, and of course this is more likely to arise during high temperature operation. In general, it's essential to have separate heaters for the sample and for the indenter tip, to ensure that they are at the same temperature when they are brought into contact. Furthermore, in order for a true steady state to be established prior to indentation, it's often necessary to have cooling systems, as well as heaters. There is also potential interest in operating at ***cryogenic*** temperatures, although there has been much less work in this area [26, 27], partly because of condensation problems in air.

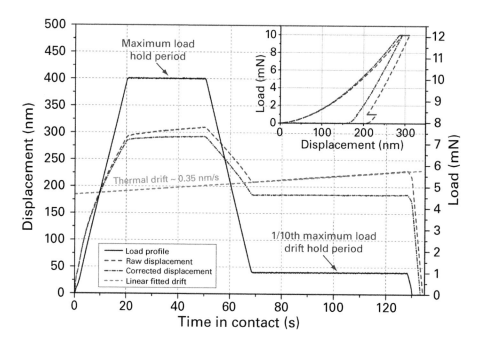

Fig. 9.4 Illustration of a thermal drift correction procedure [25].

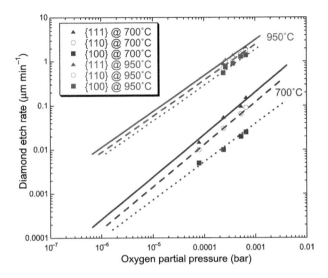

Fig. 9.5 Oxidative etch rates [28] for diamond (low energy) crystallographic faces, as a function of oxygen partial pressure and temperature, with power law extrapolations.

The other major difficulty associated with high temperature operation concerns the danger of oxidation or other chemical reaction, either of the sample or of the indenter tip (or between the two). As mentioned above in §9.2.2, choice of the material for the indenter may be different if high temperature operation is envisaged. In particular, use of diamond (in air) is likely to be problematic. This is illustrated by the diamond oxidation rate data [28] shown in Fig. 9.5. It can be seen that, even at 700 °C, which is a relatively moderate temperature for many purposes, the rate of oxidation is around 0.1 μm per minute at an oxygen partial pressure of 1 mbar. In air (1 bar), the corresponding rate would be in the micron per second range. Furthermore, this rate would tend to be even higher at edges and apices. Clearly, a diamond tip would become heavily eroded immediately under such conditions. In practice, diamond cannot be used in air at temperatures above about 300–400 °C: as described in §9.2.2, an alternative such as sapphire would be necessary.

It is also clear from Fig. 9.5 that the problem becomes much less severe at low oxygen partial pressures. With a value around 10^{-6} bar, which is routinely achievable in a vacuum system, oxidation rates, even at 950 °C, are in the nm min^{-1} range, which may well be acceptable. Furthermore, oxidation of the sample is also likely to be negligible in such atmospheres. Moreover, if the system is located in a vacuum chamber, then a reducing atmosphere can be introduced, so that there may be no real need for a high vacuum (which, due to outgassing effects, is in any event unlikely to be achieved in the presence of a heating system).

Nanoindenters certainly can be housed within vacuum systems and an example of this is shown [29] in Fig. 9.6. This is not entirely simple, particularly if the indenter was not originally designed to operate under vacuum, but customized commercial systems are available. It may also be noted that there has been an increasing tendency

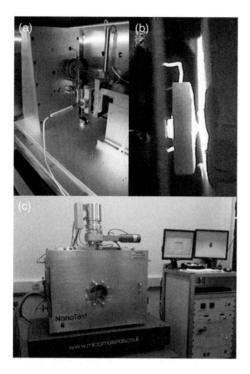

Fig. 9.6 Photos [29] of a vacuum-housed nanoindentation system, showing (a) the (pendulum-based) loading system, (b) a sample at 950 °C and (c) an external view of the vacuum chamber.

over recent years to carry out nanoindentation, and similar high precision mechanical testing operations, within the vacuum chambers of electron microscopes (SEMs [30, 31] and TEMs [20, 32]). Of course, this brings the advantage of a potential for *in situ* imaging of the sample during loading, as well as the provision of a vacuum, although it also tends to limit the size of the sample and the geometry and load capacity of the system. Nevertheless, it is now quite widely used for nanoindentation and, particularly, for micropillar compression – see §9.5.

9.3 Nanoindentation Testing Outcomes

9.3.1 Background

Nanoindentation has been very widely used over recent decades, mainly with the objective of mechanically characterizing small volumes of material – typically with dimensions of the order of a few microns or less. As mentioned at several other points in this book, such as in §5.1.3 and §8.2.1, this immediately raises the issue of whether the response of such regions can reflect the bulk properties of the material (irrespective of how the testing is done). For polycrystalline materials, this is unlikely (unless the grain size is very fine), since only an assembly of a number of grains will respond in a way that is closely representative of the bulk. (It would be expected that the required volume would be much smaller for an amorphous material, but of course **glassy metals** are rather unusual.)

This is a fundamental barrier to the concept of using nanoindentation to probe local mechanical characteristics (of metals), and it applies equally to elastic and plastic properties, although the question naturally arises as to the likely magnitude of the errors associated with this effect. In fact, this is not a simple matter, since errors can also arise from other effects associated with testing on a fine scale, such as those due to surface roughness, oxide or other extraneous surface films, fine scale tip defects etc. Of course, probing the mechanical nature of fine scale features such as precipitates or thin coatings may be of interest, even if the absolute values of extracted properties are not very accurate. In any event, a brief outline is given below of how different types of mechanical property are commonly obtained from nanoindentation data, and the main issues associated with these procedures.

9.3.2 Measurement of Stiffness and Hardness

An early breakthrough in use of nanoindentation came with recognition that the stiffness of the tested volume can be obtained from the load–displacement plot. The Young's modulus can be evaluated from (the initial gradient of) the unloading part of the plot. At least to a good approximation, the sample undergoes only elastic deformation during this unloading. The methodology was first fully described by Oliver and Pharr [1], although it built on several earlier studies. Various refinements have subsequently been introduced [3, 33–35], but the original Oliver and Pharr procedure is briefly outlined here, based on their schematic sectional diagram and load–displacement plot, shown in Fig. 9.7. Their contributions included noting that the unloading curve is non-linear and that corrections are needed to account for this

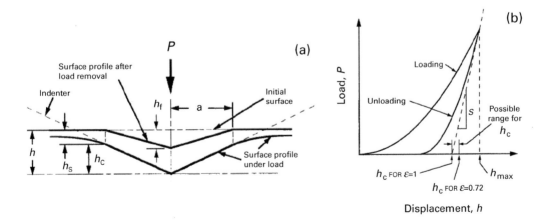

Fig. 9.7 Schematic representations [1] of: (a) a section through the sample during and after indentation and (b) a load–displacement plot.

non-linearity, for the shape of the indenter (and hence the contact area as a function of depth – sometimes termed the "*area function*" – see §7.1 – and the projected area) and also for the frame compliance. In addition, as recognized previously, the stiffness of the indenter has an effect, which is usually handled in the form of a "*reduced modulus*," E_r, incorporating E and ν for both indenter and sample:

$$E_r = \frac{(1-\nu^2)}{E} + \frac{(1-\nu_i^2)}{E_i} \tag{9.1}$$

where the subscript i refers to the indenter. The expression for this reduced modulus, obtained from nanoindentation data, is often written as

$$E_r = \frac{1}{\varepsilon} \frac{\sqrt{\pi}}{2} \frac{S}{\sqrt{A_p(h_c)}} \tag{9.2}$$

where S is the gradient shown in Fig. 9.7, $A_p(h_c)$ is the projected area at a contact depth of h_c and ε is a dimensionless constant with a value of the order of unity (see Fig. 9.7). An expression for A_p, as a polynomial in h_c, is established for each indenter shape, as is the value of ε.

Among the complications is that the profile shape may not be as shown in Fig. 9.7, incorporating "sink-in," but rather exhibit "pile-up" around the indent. In fact, as described in Chapter 8 (e.g. Fig. 8.10), this is common. It was recognized in later papers [36, 37] that this can change the contact area by factors of around 50%, introducing large errors into the inferred stiffness value. The formulation is therefore far from being a simple one, but it is often incorporated into control software supplied with commercial nanoindenters, so it is widely used. In fact, as a practical operation, it is much easier than carrying out a conventional tensile test for stiffness measurement, which usually requires a strain gauge on the sample (§5.2.2), although there is no doubt that the latter is much more likely to give a reliable and accurate result.

In fact, the procedure does usually lead to a reasonably reliable value for the Young's modulus of the sample, but a few points should be noted. Firstly, the Poisson ratio needs to be known. This is not much of an issue, since the sensitivity of the outcome to this value is relatively low, ν tends to fall within a fairly narrow range and it's usually possible to at least make a fairly reliable estimate of its value. More of a concern is that, since in many cases the region being tested is a single crystal, the fact that the Young's modulus and Poisson ratio both vary with crystallographic direction is relevant (but is normally ignored in the extraction of E). This variation[1] in Young's modulus [38] is small in some cases – zero for W and a factor of about 20% between highest and lowest for Al – but in other cases it is large – a factor of about 2.3 for Ni and 2.9 for Cu. It's rather unclear whether the value of E obtained via nanoindentation is in fact expected to relate to a particular direction, such as the

---

[1] The data provided in (Chapter VIII of) Nye for selected cubic crystals is in the form of the values of the compliances S_{11}, S_{12} and S_{44}, together with an expression for the Young's modulus in terms of them: the highest stiffness is normally in the [111] direction, while the lowest is in the [100] direction.

normal to the free surface, but in any event only a single number is obtained. Possibly it is effectively a "direction-averaged" value, in which case it might be expected to be close to that for the corresponding untextured polycrystal, but this issue is rarely addressed. In practice, the value obtained probably can't be treated as sufficiently accurate for this to be much of an issue. In any event, measurement of E is a useful capability, although it would be relatively unusual for it to be a key measured property – it could in most cases be estimated solely on the basis of the composition (and phase constitution) of the region concerned.

The measurement of hardness via the load–displacement plot, which is also commonly done, should be regarded rather differently. Depending on exactly how the hardness number is defined, then a value can certainly be obtained. The most common definition – as detailed in §7.1 – is simply the applied force divided by the contact area. Since the procedure for obtaining E involves assessment of the contact area (as a function of load and penetration depth), it is straightforward to evaluate the hardness. Of course, as explained in Chapter 7, since, even in the "macro" regime, this number varies with the applied load, and with the indenter shape, it is not well defined or rigorously related to the plasticity characteristics of the sample. It's therefore not really possible to say whether a derived value is "reliable," since there is no absolute yardstick against which it can be compared. There have in fact been many attempts [39, 40] to obtain genuine plasticity characteristics from nanoindentation data – mostly the yield stress, but also in some cases the work hardening characteristics. However, to say the least, these are not reliable procedures. This is clear just from the strong "size effects" that are observed – i.e. a dependence of the values obtained on the size of the indent – which are described in §9.4.

9.3.3 Characterization of Creep

A number of analytical or semi-analytical procedures [41–49] have been developed for obtaining creep characteristics from (depth-sensing) nanoindentation data, although they are all based on a similar approach. The focus is usually on the "secondary" creep regime (§3.2.4), in which the creep strain rate is taken as constant and given by

$$\frac{d\varepsilon_{cr}}{dt} = C\sigma^n e^{-Q/(RT)} \qquad (9.3)$$

where C is a constant, σ is the (uniaxial) stress, n is the stress exponent, T is the absolute temperature and Q is an activation energy. During indentation (creep), the stress and strain fields vary with position and time, and the behavior tends to be dominated by the primary creep regime, but it has been common to use the following version of Eqn. (9.3):

$$\frac{1}{h}\frac{dh}{dt} = C\left(\frac{P}{A_p(h)}\right)^n e^{-Q/(RT)} \qquad (9.4)$$

where h is the indentation depth, (dh/dt) is the velocity of the indenter (during a "creep dwell" period), P is the indenter load (held constant during the "creep dwell" period) and $A_p(h)$ is the projected contact area between indenter and specimen (as a function of h). It's quite common for a plot of indentation depth against time to exhibit an approximately constant gradient (penetration velocity), after a short initial transient, in a way that is reminiscent of a uniaxial creep strain plot. Thus, as with conventional (uniaxial) creep testing, a value for the stress exponent can be obtained as the gradient of a plot of (the log of) the steady state creep strain rate against (the log of) the stress. However, use of Eqn. (9.4) in this way is based on the following assumptions:

1. At any time, the stress field beneath the indenter can be represented by a single "equivalent" value (given by the load over the current projected contact area).
2. At any time, the changing strain field beneath the indenter can be represented by a single "indentation creep strain rate" (given by the current indenter velocity over the current depth).
3. Once the indenter velocity has become (approximately) constant, "secondary" creep is fully established throughout all the parts of the sample affecting the indenter penetration (and primary creep can be ignored).

In fact, it has been shown [50–53] that none of these assumptions is justifiable and the value obtained for n via this methodology can vary widely, depending on indenter size, applied load, test period and details of the data processing, so the procedure has been largely discredited. There is, of course, the additional issue of whether the response of a very small volume of material can be representative of the bulk. In practice, while it is possible to obtain information about both primary and secondary regimes of creep via inverse iterative FEM simulation of the indentation creep process, as described in §8.4.4, this "size effect" issue is such that this is not really viable using a nanoindenter. There is also the further complication that the "thermal drift" effect described in §9.2.3 is much more problematic with nanoindenters than with (coarser scale) indentation creep plastometers.

9.4 The Nanoindentation Size Effect

9.4.1 Experimental Observations

It quickly became clear during early work that the values of properties related to plastic deformation, when obtained via nanoindentation load–displacement plots, tend to vary with the size of the indent. In general, there is a clear tendency for the material to appear to become harder as the scale of the indentation is reduced. This is often referred to simply as a "*size effect*" [2, 4, 7]. The easiest parameter to obtain, although not really a "genuine" property, is the hardness number, H (usually defined as the load over the contact area – see §7.1 for an explanation of the meaning of hardness). The kind of effect that is commonly observed is shown [4] in Fig. 9.8, which gives measured hardness values, as a function of penetration depth, for single crystals of silver and gold. Although obtained by several different researchers, these data should be free of extraneous variations, since they all refer to high purity single crystals with

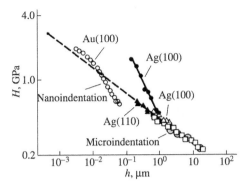

Fig. 9.8 Dependence [4] of measured Vickers hardness on indent depth, for single crystals of silver and gold. (The different symbols relate to work by different authors.)

defined orientations. Moreover, these metals should be free of significant surface oxide – certainly the gold (which forms no oxide) and probably the silver (since the Ag data refer to deeper indents).

It is therefore clear that these variations – for example, a rise from ~0.5 GPa to ~2 GPa as the indent depth is reduced from ~100 nm to ~5 nm for the Au – are "*scale-affected.*" In fact, even 0.5 GPa is a major overestimate compared with the "correct" bulk value for gold. (This should probably be around 0.1 GPa, based on a "rule of thumb" that the Vickers hardness is about $3\sigma_Y$ (Eqn. (7.5)) and the yield stress of a single crystal of pure gold is certainly no greater than a few tens of MPa, with a minor dependence on orientation.) Actually, such trends are observed for virtually all materials, and are often highly significant over the range of indent sizes commonly created during nanoindentation. There can be no question that plasticity-related properties obtained in this way are certainly not "correct" in the sense that they do not reflect an inherent (scale-independent) characteristic of the material being tested.

As can be seen in Fig. 9.8, such data often appear to follow a (power law) functional relationship [4], along the lines of H being proportional to h^{-n}, with the exponent n lying in the approximate range 0.1–0.3. This is certainly not rigorous or derivable, but it does give a feel for the sensitivity. In practice, the variations are more significant in the "nano" regime (sub-micron depths) than in the range labelled in Fig. 9.8 as "microindentation" (depths of a few microns to a few tens of microns) and tend to become small for depths above about 100 μm. On the other hand, as described in §7.2, measured hardness values can vary with load (depth) even in the "macro" hardness testing regime, although the sense of the change tends to reverse – i.e. the hardness tends to rise slightly with increasing load (depending on the work hardening characteristics).

9.4.2 Size Effect Mechanisms and "Pop-in" Phenomena

In any event, even after accepting the complications associated with the definition of hardness, it's clear that values of H obtained via nanoindentation, and also any similarly inferred values of yield stress and work hardening rates, are likely to be (gross) overestimates of the "real" values for the material concerned. Moreover, attempts [39, 40] to "correct" for the size effect are always going to be difficult without a full understanding of its origin. Various possible explanations for the effect have been put forward, many of them based on dislocation-related phenomena. For example, the concept has frequently been invoked [54–57] of a high gradient of plastic strain being associated with (fine scale) indentation, bringing a requirement for "*geometrically necessary dislocations*," which can only be created if the stress is raised. In fact, the concept of "*strain gradient plasticity*" has been widely promoted more generally [58, 59]. In the context of nanoindentation, there has been quite extensive theoretical analysis of how strain gradients, and an associated high dislocation density in certain regions, can raise the effective hardness, essentially via a similar mechanism to that operative during conventional strain hardening – i.e. the formation of various tangles, sessile jogs etc. (see §4.2.2). However, despite the many attempts at theoretical justification, it is actually quite difficult to imagine "forest hardening" mechanisms of this type being operative in the very small volumes concerned, or causing the very large size effects that are commonly observed. Furthermore, many of the models developed in the area are effectively just curve-fitting operations. Nevertheless, the idea has been persistent in some circles and of course there may be a contribution to the effect via a mechanism of this type.

However, a rather different dislocation-related mechanism that could be at least partly responsible for the nanoindentation "size effect" involves the possibility that the deformed volume initially contains no dislocations, creating a requirement for a higher stress in order to nucleate them and initiate plastic deformation. This would be needed in order to activate a new source of dislocations, either internal (such as a Frank–Read source – see §4.2.3) or at a free surface. This is plausible in the sense that, for a metal, the dislocation density, ρ, may be as low as $\sim 10^{10}$ m^{-2}, corresponding to an average distance between dislocations of ~ 10 μm ($\sim 1/\sqrt{\rho}$; see §4.2.2). Non-metals could have even greater distances between adjacent dislocations. For an indent depth of, say, a few tens of nm, the volume being plastically deformed is likely to be below a few cubic microns, and hence be effectively free of dislocations initially.

Information that supports the "*dislocation nucleation*" argument is the common observation in load–displacement plots of what is often termed a "*pop-in*" phenomenon – that is, a burst of straining while the load remains unchanged [6, 60–63]. Such an effect might certainly be expected if there is a barrier to the creation of dislocations, but they are readily able to glide once they are formed. Examples of curves [6, 61] exhibiting strong pop-in effects are shown in Fig. 9.9. Furthermore, it has been found [61] that the effect is reduced when the initial dislocation density is higher (as a result of prior cold work). This is illustrated by the data in Fig. 9.10. It certainly seem likely that this type of phenomenon can occur, although exactly what effect it is likely to have on a derived hardness (or yield stress) value will depend on how the data are processed.

It should, however, be mentioned that pop-in features in load–displacement plots may in some cases be due to effects not associated with dislocations. For example,

9.4 The Nanoindentation Size Effect

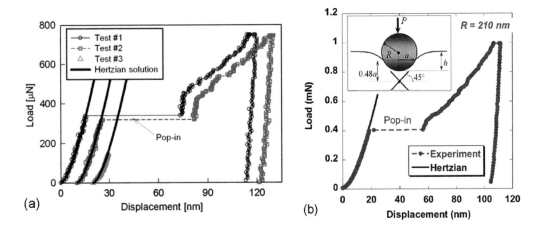

Fig. 9.9 (a) Experimental load–displacement plots [61] from nanoindentation of annealed [100] Ni single crystals (using 0.58 μm radius spheres), with the Herzian (elastic) lines superimposed, and (b) a schematic representation [6] of this type of indentation, showing the locations of the planes of peak shear stress.

(martensitic) phase transformations (see §4.4.2) stimulated by the applied load can create similar effects [64]. Moreover, while this is only rarely identified specifically as a cause of pop-in, it's probable that in a number of cases it represents some kind of break-through of a surface oxide film or other extraneous layer. Native oxide films on metals are in general relatively thin (sub-micron and often only a few nm), but they are usually much harder than the metal and, when indentation is only being carried out to a depth of a few tens of nm, their rupture is quite likely to cause a noticeable burst in a load–displacement plot. While attempts to quantify such an effect are rare, it has been recognized [65] that pop-in effects are sensitive to the details of how the near-surface region has been prepared. In any event, such effects can create further complications when attempting to characterize the real mechanical characteristics of a sample.

In summary, while the size effect is undoubtedly real, and certainly means that plasticity-related characteristics obtained via nanoindentation are unlikely to be reliable indicators of anything inherent in the material being tested, the explanation for the effect remains rather unclear. Possibly it arises from more than one mechanism, and in different combinations in individual tests. It should perhaps also be mentioned that it could be at least partly due to what might be termed "*constraint*" effects. Some of the ways in which such effects can influence test outcomes are described in Chapter 5, mostly on a more macroscopic scale. During nanoindentation, the deformation of the region being tested is constrained by it being bonded to surrounding regions that remain rigid, or at least elastic. The nature and strength of such effects will vary with the indenter geometry and material properties, but they will certainly tend to become more significant as the volume of the region being tested becomes smaller. Unfortunately, this is complex geometrically, and also not well suited to treatment of the region as an isotropic continuum, so it is far from easy to analyze or validate, and

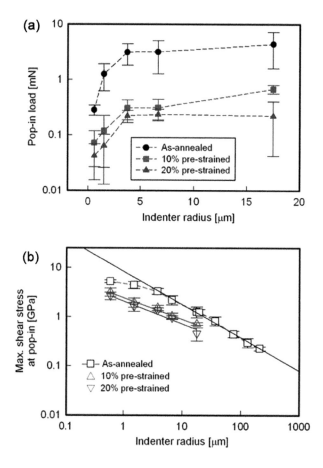

Fig. 9.10 Data [61] illustrating the effect of prior cold work on the pop-in behavior of [100] Ni single crystals, showing (a) measured pop-in loads and (b) peak shear stresses (determined from pop-in loads), as a function of indenter radius.

indeed this has rarely been attempted. The concept does have some aspects in common with that of strain gradient plasticity, but quantification, on any basis, presents a major challenge. Overall, it has to be recognized that "correcting" in a well-defined way for the size effect is not really a viable proposition at present, and indeed it looks likely to remain a rather intractable problem.

9.5 Micropillar Compression Testing

9.5.1 Creation of Micropillar Samples

The idea of carrying out compression testing on small regions effectively arose simply from the availability of nanoindenters – i.e. of a capability for applying small loads in a controlled way and accurately monitoring the displacement as the load is increased.

In fact, this is not the only type of "spin-off" testing that has arisen, since equipment of this type has also been used for "push-out" of fibers in composite materials [66–69], aimed at obtaining information about mechanical characteristics of the interface. Compression testing on a fine scale offers the attraction of creating much simpler stress and strain fields than those arising during indentation, and hence the potential for easier analysis of the output data, while still being able to probe small volumes of material.

Of course, if nanoindenters, or set-ups with capabilities of a similar type, are to be used, then there are limitations to the dimensions of the sample. Taking the maximum load capability to be ~1 N (a relatively high value), then, if samples with yield stress levels of up to, say, 500 MPa are to be deformed plastically, then the maximum sample diameter is about 50 μm. In practice, micropillars of up to around this diameter are tested, but most are smaller. Of course, this also introduces a limit on the sample length, since, as outlined in §6.1.2, the aspect ratio of compression samples must be kept below about 2 or 3 if buckling is to be avoided. Samples with dimensions such as these clearly cannot be handled and manipulated in the same way as for conventional compression testing. The test can only be carried out if the region to be tested is somehow secured at the base and held vertical (normal to the loading axis). In practice, this can only be done via some sort of machining or preferential dissolution of the sample *in situ*, leaving the base integral with the substrate.

Certain techniques that are quite well suited to such fine scale shaping were available by the time that this type of testing was being initiated (around the early 2000s). One possibility is lithographic procedures. Lithography is a very old technology, with a long history in the printing industry, but it was brought to a high level of sophistication in the latter part of the twentieth century, largely for production of electronic integrated circuits, involving various masking and dissolution processes. In fact, there are several different variants of the process, involving light (visible or UV) or electron beams. An alternative, which was developed later, is *focussed ion beam* (FIB) milling. This is a rather slow and expensive process, but it is very versatile in terms of shape capability (and can be used to create more complex sample shapes than cylinders, including mini "tensile" samples [70]). Discussions of the pros and cons of the two approaches, for micropillar compression purposes, are available in the literature [71, 72].

The FIB milling procedure has been widely used, commonly with software controlling the movement of the ion beam. Various ions have been used, but most production is carried out using Ga^+ ions, usually with accelerating voltages of a few tens of kV and currents in the nA range. Technical details are available in the literature [70, 73]. Typical products [70] are shown in Fig. 9.11, where it can be seen that good dimensional control is possible, even when the dimensions of the pillar are of the order of a few microns. This procedure doesn't allow the testing of the very small (submicron) regions that can, at least in principle, be interrogated via nanoindentation, but it does nevertheless offer good spatial resolution. As with nanoindentation, the regions being tested will commonly be single crystals (with the associated advantages and disadvantages). The compressive load is normally applied using a system akin to those

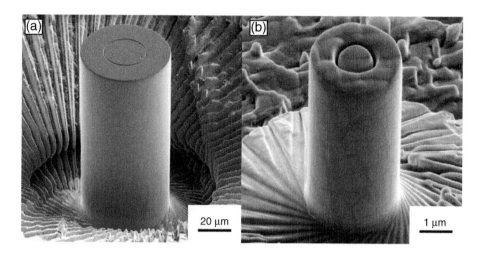

Fig. 9.11 SEM micrographs [70] of two Ni-base superalloy micropillars produced by FIB milling, with dimensions of (a) 43 μm diameter and 90 μm height and (b) 2.3 μm diameter and 4.6 μm height (with Ni_3Al precipitates visible on the surface). The milled circular marks on the top surface were created for guidance purposes.

employed for nanoindentation, via a flat-ended punch. There is, of course, a challenge involved in ensuring that the axes of indenter and sample are lined up reasonably well. In practice, this is often facilitated by the operation being carried out inside the chamber of an SEM microscope.

9.5.2 Test Issues

It may be noted at this point that there are at least a couple of major concerns about this type of test procedure. One is that the constraint conditions at the bottom of the column, where the sample is continuous with the substrate, are different from those in conventional compressive testing. Clearly, there can be no frictional sliding there, which commonly occurs in macroscopic testing (§6.2.2). Another problem, which is limited to samples produced by FIB, is that the ion beam milling tends to leave ions implanted in the sample (typically [74, 75] to a depth of at least several nm). It is now clear [71, 73] that this can affect the plasticity, particularly with relatively small samples, with a tendency to inhibit dislocation glide (and also to reduce the toughness). The type of effect that arises is illustrated by the data [75] shown in Fig. 9.12. In this work, the (LiF) samples were created either by FIB milling of a LiF substrate or by dissolution of the NaCl constituent of a LiF-NaCl directionally grown rod-like eutectic (with the diameter of the LiF rods being controlled via the eutectic growth velocity). It is clear that implantation of the Ga^+ ions during the FIB has significantly affected the mechanical response, although, over this limited range there is not much evidence of a size effect. It is possible to produce micropillar samples in other ways, such as via lithographic techniques, or possibly by differential dissolution in special

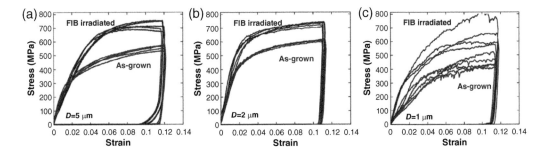

Fig. 9.12 Collections of stress–strain curves [75] from micropillars of LiF, created either by FIB or by a dissolution technique and compressed along <111>, for groups with nominal diameters of (a) 5 μm, (b) 2 μm and (c) 1 μm.

cases such as that described above, but these tend to be less versatile and have been less frequently used.

9.5.3 Test Outcomes and Size Effects

Testing is normally carried out well into the plastic regime. Indeed, the main potential for useful experiments lies in detailed study of the plasticity characteristics of small volumes. Some examples [70] of the appearance of samples after testing are shown in Fig. 9.13. These are all samples of the same superalloy, with the only difference between them being one of size (diameter). These were all milled from the same substrate (having a <269> direction normal to the free surface). There was no control over the direction from which they were being viewed in these micrographs, but the same primary slip system (§4.2.1) came into operation first for each sample. (This is the one on which the shear stress is largest, which in this case is the (–111)[101] system: using Eqn. (3.5), the corresponding Schmid factor can be evaluated as about 0.482.) For samples (B)–(D), which were subjected to relatively low strains of ~2–4%, only this slip system operated, as in the schematic depiction of Fig. 4.10(c). For sample (A), however, which was strained to over 8%, it can be seen that a second slip system came into operation.

Of course, it would be hoped that the stress–strain curves obtained from these tests would all correspond to the outcome of a conventional compressive test on the same single crystal (in the same orientation), although clearly the requirements concerning the measurement accuracy for both the sample diameter and the displacement are demanding for the micropillars. In reality, the outcomes [70], which are shown in Fig. 9.14 for sets of samples collected together by diameter, are rather different. Firstly, there is some scatter – rather more than might have been expected for conventional testing. Secondly, there is a clear tendency for the apparent yield stress to be higher for the smaller samples. Furthermore, the values are all rather higher than for corresponding conventional testing, which is quoted [70] as about 850 MPa. Such a size effect is reminiscent of nanoindentation outcomes, and, as in that case, is almost

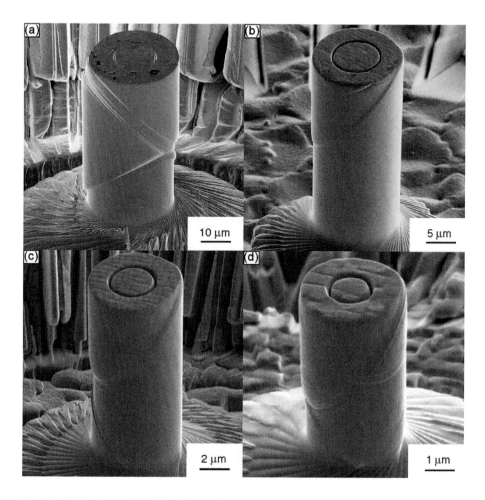

Fig. 9.13 SEM micrographs [70] of four Ni-base superalloy (UM-F19) micropillars after compressive testing, having (a) 20.6 μm diameter (nominal plastic strain 8.5%), (b) 9.4 μm diameter (2.1% strain), (c) 4.8 μm diameter (4% strain) and (d) 2.3 μm diameter (3.5% strain).

universally observed – see §9.4.2. Finally, the shapes of the curves are not entirely as might have been expected for the higher strain samples, since the onset of multiple slip is usually reflected in an increasing flow stress (Fig. 4.11).

Errors of this type are of concern, since the main attraction of this testing is a capability to study basic plasticity characteristics of small regions, with the added attraction of a potential for viewing features such as individual persistent slip bands (Fig. 3.32) on the surface, which is normally difficult during conventional testing – although it can be done after indentation plastometry (Fig. 7.12). However, if measured yield stress values, and changes in flow stress, are being affected by sample preparation artefacts and size effects, then their usefulness is clearly reduced. These size effects tend to cover a smaller range than those typically observed during nanoindentation, although the size range itself is usually also smaller. Some typical

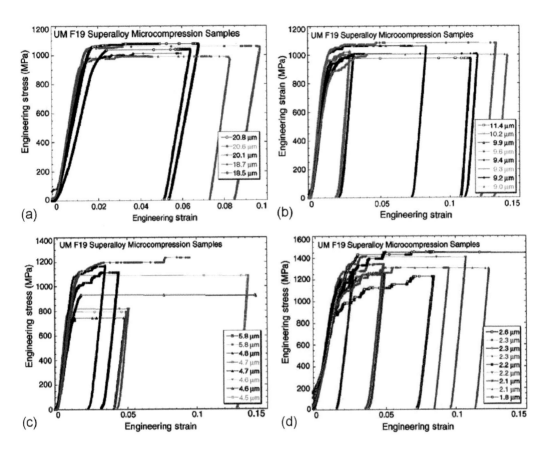

Fig. 9.14 Collections of stress–strain curves [70] from micropillars of the type shown in Fig. 9.13, for groups with nominal diameters of (a) 20 μm, (b) 10 μm, (c) 5 μm and (d) 2 μm.

size effect data [76] are shown in Fig. 9.15, where a clear trend is apparent. Several reviews [8, 13–15] are available covering this issue. In general, similar explanations to those invoked for nanoindentation tend to be put forward, including strain gradient arguments (geometrically necessary dislocations) and dislocation nucleation barriers (although this is only likely at the low end of the testable size range). The "constraint" argument (which is rarely expressed in either case using that term) cannot operate in the same way, since most of the sample is clearly unconstrained. On the other hand, the fact that the tested region is continuous with the substrate at its base clearly exerts a significant constraint on the deformation, perhaps depending slightly on the sample height.

A general conclusion regarding this type of testing has to be that, while outcomes may be interesting, and attractive images are sometimes created, it suffers from similar drawbacks to nanoindentation in the sense that, as a consequence of various extraneous effects that are difficult to quantify, it cannot be used to obtain reliable, or even meaningful, information about the real mechanical (plasticity-related) characteristics of the region being tested.

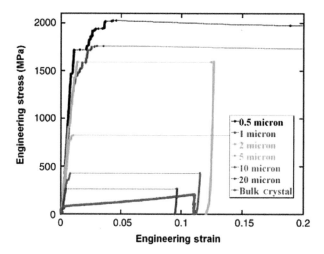

Fig. 9.15 Representative stress–strain curves [76] from micropillars of Ni_3Al-Ta, with <123> orientation, over a range of pillar diameters.

References

1. Oliver, WC and GM Pharr, An improved technique for determining hardness and elastic-modulus using load and displacement sensing indentation experiments. *Journal of Materials Research*, 1992. **7**(6): 1564–1583.
2. Wei, YG, XZ Wang and MH Zhao, Size effect measurement and characterization in nanoindentation test. *Journal of Materials Research*, 2004. **19**(1): 208–217.
3. Fischer-Cripps, AC, Critical review of analysis and interpretation of nanoindentation test data. *Surface and Coatings Technology*, 2006. **200**: 4153–4165.
4. Golovin, Y, Nanoindentation and mechanical properties of solids in submicrovolumes, thin near-surface layers, and films: a review. *Physics of the Solid State*, 2008. **50**(12): 2205–2236.
5. Oliver, WC and GM Pharr, Nanoindentation in materials research: past, present, and future. *MRS Bulletin*, 2010. **35**(11): 897–907.
6. Morris, JR, H Bei, GM Pharr and EP George, Size effects and stochastic behavior of nanoindentation pop in. *Physical Review Letters*, 2011. **106**(16).
7. Voyiadjis, GZ and M Yaghoobi, Review of nanoindentation size effect: experiments and atomistic simulation. *Crystals*, 2017. **7**(10).
8. Shahbeyk, S, GZ Voyiadjis, V Habibi, SH Astaneh and M Yaghoobi, Review of size effects during micropillar compression test: experiments and atomistic simulations. *Crystals*, 2019. **9**(11).
9. Fischer-Cripps, AC, *Nanoindentation*. Mechanical Engineering Series, Ling, FF, ed. New York: Springer-Verlag, 2004.
10. Chen, L, ed. *Micro-Nanoindentation in Materials Science*. ML Books International, 2015.
11. Tiwari, A and S Natarajan, eds. *Applied Nanoindentation in Advanced Materials*. Hoboken, NJ: Wiley, 2017.
12. Tsui, T and AA Volinsky, eds. *Small Scale Deformation Using Advanced Nanoindentation Techniques*. MDPI, 2019.

13. Soler, R, JM Wheeler, HJ Chang, J Segurado, J Michler, J Llorca and JM Molina-Aldareguia, Understanding size effects on the strength of single crystals through high-temperature micropillar compression. *Acta Materialia*, 2014. **81**: 50–57.
14. Bittencourt, E, Interpretation of the size effects in micropillar compression by a strain gradient crystal plasticity theory. *International Journal of Plasticity*, 2019. **116**: 280–296.
15. Takata, N, S Takeyasu, HM Li, A Suzuki and M Kobashi, Anomalous size-dependent strength in micropillar compression deformation of commercial-purity aluminum single-crystals. *Materials Science and Engineering: A Structural Materials: Properties, Microstructure and Processing*, 2020. **772**.
16. Li, WD, HB Bei, J Qu and YF Gao, Effects of machine stiffness on the loading–displacement curve during spherical nano-indentation. *Journal of Materials Research*, 2013. **28**(14): 1903–1911.
17. Dargenton, JC and J Woirgard, Description of an electrostatic force nanoindenter. *Journal de Physique III*, 1996. **6**(9): 1247–1260.
18. Woirgard, J and JC Dargenton, A new proposal for design of high accuracy nanoindenters. *Measurement Science and Technology*, 1995. **6**(1): 16–21.
19. Yu, N, WA Bonin and AA Polycarpou, High-resolution capacitive load-displacement transducer and its application in nanoindentation and adhesion force measurements. *Review of Scientific Instruments*, 2005. **76**(4).
20. Bobji, MS, CS Ramanujan, JB Pethica and BJ Inkson, A miniaturized TEM nanoindenter for studying material deformation in situ. *Measurement Science and Technology*, 2006. **17**(6): 1324–1329.
21. Elhebeary, M and MTA Saif, A micromechanical bending stage for studying mechanical properties of materials using nanoindenter. *Journal of Applied Mechanics: Transactions of the ASME*, 2015. **82**(12).
22. Wheeler, JM and J Michler, Invited article: indenter materials for high temperature nanoindentation. *Review of Scientific Instruments*, 2013. **84**(10): 101301.
23. Monclus, MA, S Lotfian and JM Molina-Aldareguia, Tip shape effect on hot nanoindentation hardness and modulus measurements. *International Journal of Precision Engineering and Manufacturing*, 2014. **15**(8): 1513–1519.
24. Nohava, J, NX Randall and N Conte, Novel ultra nanoindentation method with extremely low thermal drift: principle and experimental results. *Journal of Materials Research*, 2009. **24**(3): 873–882.
25. Wheeler, JM, DEJ Armstrong, W Heinz and R Schwaiger, High temperature nanoindentation: the state of the art and future challenges. *Current Opinion in Solid State & Materials Science*, 2015. **19**(6): 354–366.
26. Chen, J, GA Bell, HS Dong, JF Smith and BD Beake, A study of low temperature mechanical properties and creep behaviour of polypropylene using a new sub-ambient temperature nanoindentation test platform. *Journal of Physics D: Applied Physics*, 2010. **43**(42).
27. Lee, SW, YT Cheng, I Ryu and JR Greer, Cold-temperature deformation of nano-sized tungsten and niobium as revealed by in-situ nano-mechanical experiments. *Science China Technological Sciences*, 2014. **57**(4): 652–662.
28. Wheeler, JM, RA Oliver and TW Clyne, AFM observation of diamond indenters after oxidation at elevated temperatures. *Diamond & Related Materials*, 2010. **19**: 1348–1353.
29. Harris, A, BD Beake, DEJ Armstrong and MI Davies, Development of high temperature nanoindentation methodology and its application in the nanoindentation of polycrystalline tungsten in vacuum to 950°C. *Experimental Mechanics*, 2017. **57**(7): 1115–1126.

30. Rzepiejewska-Malyska, KA, G Buerki, J Michler, RC Major, E Cyrankowski, SAS Asif and OL Warren, In situ mechanical observations during nanoindentation inside a high-resolution scanning electron microscope. *Journal of Materials Research*, 2008. **23**(7): 1973–1979.
31. Nowak, JD, KA Rzepiejewska-Malyska, RC Major, OL Warren and J Michler, In-situ nanoindentation in the SEM. *Materials Today*, 2010. **12**: 44–45.
32. Warren, OL, ZW Shan, SAS Asif, EA Stach, JW Morris and AM Minor, In situ nanoindentation in the TEM. *Materials Today*, 2007. **10**(4): 59–60.
33. Li, XD and B Bhushan, A review of nanoindentation continuous stiffness measurement technique and its applications. *Materials Characterization*, 2002. **48**(1): 11–36.
34. Wang, LG and SI Rokhlin, Universal scaling functions for continuous stiffness nanoindentation with sharp indenters. *International Journal of Solids and Structures*, 2005. **42**(13): 3807–3832.
35. Pharr, GM, JH Strader and WC Oliver, Critical issues in making small-depth mechanical property measurements by nanoindentation with continuous stiffness measurement. *Journal of Materials Research*, 2009. **24**(3): 653–666.
36. Bolshakov, A and GM Pharr, Influences of pileup on the measurement of mechanical properties by load and depth sensing indentation. *Journal of Material Research*, 1998. **13**(4): 1049–1058.
37. Hay, JC, A Bolshakov and GM Pharr, A critical examination of the fundamental relations used in the analysis of nanoindentation data. *Journal of Materials Research*, 1999. **14**(6): 2296–2305.
38. Nye, JF, *Physical Properties of Crystals – Their Representation by Tensors and Matrices*. Oxford: Clarendon, 1985.
39. Kim, SH, YC Kim, S Lee and JY Kim, Evaluation of tensile stress–strain curve of electroplated copper film by characterizing indentation size effect with a single nanoindentation. *Metals and Materials International*, 2017. **23**(1): 76–81.
40. Chen, X, IA Ashcroft, RD Wildman and CJ Tuck, A combined inverse finite element – elastoplastic modelling method to simulate the size-effect in nanoindentation and characterise materials from the nano to micro-scale. *International Journal of Solids and Structures*, 2017. **104**: 25–34.
41. Liu, H, Y Chen, Y Tang, S Wei and G Nuiu, Tensile and indentation creep behaviour of Mg-5%Sn and Mg-5%Sn-2%Di alloys. *Materials Science and Engineering A*, 2007. **464**: 124–128.
42. Takagi, H, M Dao and M Fujiwara, Analysis on pseudo-steady indentation creep. *Acta Mechanica Solida Sinica*, 2008. **21**: 283–288.
43. Marques, VMF, B Wunderle, C Johnston and PS Grant, Nanomechanical characterisation of Sn-Ag-Cu/Cu joints – part 2: nanoindentation creep and its relationship with uniaxial creep as a function of temperature. *Acta Materialia*, 2013. **61**(7): 2471–2480.
44. Geranmayeh, AR and R Mahmudi, Indentation creep of a cast Mg-6Al-1Zn-0.7Si alloy. *Materials Science and Engineering: A Structural Materials: Properties, Microstructure and Processing*, 2014. **614**: 311–318.
45. Chatterjee, A, M Srivastava, G Sharma and JK Chakravartty, Investigations on plastic flow and creep behaviour in nano and ultrafine grain Ni by nanoindentation. *Materials Letters*, 2014. **130**: 29–31.
46. Wang, Y and J Zeng, Effects of Mn addition on the microstructure and indentation creep behaviour of the hot dip Zn coating. *Materials & Design*, 2015. **69**: 64–69.

47. Mahmudi, R, M Shalbafi, M Karami and AR Geranmayeh, Effect of Li content on the indentation creep characteristics of cast Mg-Li-Zn alloys. *Materials & Design*, 2015. **75**: 184–190.
48. Ma, Y, GJ Peng, DH Wen, and TH Zhang, Nanoindentation creep behavior in a CoCrFeCuNi high-entropy alloy film with two different structure states. *Materials Science and Engineering: A Structural Materials: Properties, Microstructure and Processing*, 2015. **621**: 111–117.
49. Ginder, RS, WD Nix and GM Pharr, A simple model for indentation creep. *Journal of the Mechanics and Physics of Solids*, 2018. **112**: 552–562.
50. Goodall, R and TW Clyne, A critical appraisal of the extraction of creep parameters from nanoindentation data obtained at room temperature. *Acta Materialia*, 2006. **54**(20): 5489–5499.
51. Chen, J and SJ Bull, The investigation of creep of electroplated Sn and Ni-Sn coating on copper at room temperature by nanoindentation. *Surface and Coatings Technology*, 2009. **203**(12): 1609–1617.
52. Dean, J, J Campbell, G Aldrich-Smith and TW Clyne, A critical assessment of the "stable indenter velocity" method for obtaining the creep stress exponent from indentation data. *Acta Materialia*, 2014. **80**: 56–66.
53. Campbell, J, J Dean and TW Clyne, Limit case analysis of the "stable indenter velocity" method for obtaining creep stress exponents from constant load indentation tests. *Mechanics of Time-dependent Materials*, 2016. **1**: 31–43.
54. Nix, WD and H Gao, Indentation size effects in crystalline materials: a law for strain gradient plasticity. *Journal of the Mechanics and Physics of Solids*, 1998. **46**: 411–425.
55. Elmustafa, AA and DS Stone, Nanoindentation and the indentation size effect: kinetics of deformation and strain gradient plasticity. *Journal of the Mechanics and Physics of Solids*, 2003. **51**: 357–381.
56. Zhao, MH, WS Slaughter, M Li and SX Mao, Material-length-scale-controlled nanoindentation size effects due to strain-gradient plasticity. *Acta Materialia*, 2003. **51**(15): 4461–4469.
57. Lee, H, S Ko, J Han, H Park and W Hwang, Novel analysis for nanoindentation size effect using strain gradient plasticity. *Scripta Materialia*, 2005. **53**(10): 1135–1139.
58. Gao, H, H Huang, WD Nix and JW Hutchinson, Mechanism-based strain gradient plasticity – I: theory. *Journal of Mechanics and Physics of Solids*, 1999. **47**: 1239–1263.
59. Fleck, NA and JW Hutchinson, A reformulation of strain gradient plasticity. *Journal of the Mechanics and Physics of Solids*, 2001. **49**(10): 2245–2271.
60. Lorenz, D, A Zeckzer, U Hilpert, P Grau, H Johansen, and HS Leipner, Pop-in effect as homogeneous nucleation of dislocations during nanoindentation. *Physical Review B*, 2003. **67**(17).
61. Shim, S, H Bei, EP George and GM Pharr, A different type of indentation size effect. *Scripta Materialia*, 2008. **59**(10): 1095–1098.
62. Barnoush, A, MT Welsch and H Vehoff, Correlation between dislocation density and pop-in phenomena in aluminum studied by nanoindentation and electron channeling contrast imaging. *Scripta Materialia*, 2010. **63**(5): 465–468.
63. Ahn, TH, CS Oh, K Lee, EP George and HN Han, Relationship between yield point phenomena and the nanoindentation pop-in behavior of steel. *Journal of Materials Research*, 2012. **27**(1): 39–44.
64. Chrobak, D, K Nordlund and R Nowak, Nondislocation origin of GaAs nanoindentation pop-in event. *Physical Review Letters*, 2007. **98**(4).

65. Wang, ZG, H Bei, EP George and GM Pharr, Influences of surface preparation on nanoindentation pop-in in single-crystal Mo. *Scripta Materialia*, 2011. **65**(6): 469–472.
66. Watson, MC and TW Clyne, The tensioned push-out test for measurement of fibre/matrix interfacial toughness under mixed mode loading. *Materials Science and Engineering*, 1993. **A160**: 1–5.
67. Eldridge, JI and BT Ebihara, Fiber push-out testing apparatus for elevated temperatures. *Journal of Materials Research*, 1994. **9**(4): 1035–1042.
68. Kalton, AF, SJ Howard, J Janczak-Rusch and TW Clyne, Measurement of interfacial fracture energy by single fibre push-out testing and its application to the titanium–silicon carbide system. *Acta Materiala*, 1998. **46**: 3175–3189.
69. Rebillat, F, J Lamon, R Naslain, E Lara-Curzio, MK Ferber and TM Besmann, Interfacial bond strength in SiC/C/SiC composite materials, as studied by single-fiber push-out tests. *Journal of the American Ceramic Society*, 1998. **81**(4): 965–978.
70. Uchic, MD and DA Dimiduk, A methodology to investigate size scale effects in crystalline plasticity using uniaxial compression testing. *Materials Science and Engineering: A Structural Materials: Properties, Microstructure and Processing*, 2005. **400**: 268–278.
71. Burek, MJ and JR Greer, Fabrication and microstructure control of nanoscale mechanical testing specimens via electron beam lithography and electroplating. *Nano Letters*, 2010. **10**(1): 69–76.
72. Chen, M, J Wehrs, J Michler and JM Wheeler, High-temperature in situ deformation of GaAs micro-pillars: lithography versus FIB machining. *JOM*, 2016. **68**(11): 2761–2767.
73. Hutsch, J and ET Lilleodden, The influence of focused-ion beam preparation technique on microcompression investigations: lathe vs. annular milling. *Scripta Materialia*, 2014. **77**: 49–51.
74. Kiener, D, C Motz, M Rester, M Jenko and G Dehm, FIB damage of Cu and possible consequences for miniaturized mechanical tests. *Materials Science and Engineering: A Structural Materials: Properties, Microstructure and Processing*, 2007. **459**(1–2): 262–272.
75. Soler, R, JM Molina-Aldareguia, J Segurado, J Llorca, RI Merino and VM Orera, Micropillar compression of LiF 111 Single crystals: effect of size, ion irradiation and misorientation. *International Journal of Plasticity*, 2012. **36**: 50–63.
76. Uchic, MD, DM Dimiduk, JN Florando and WD Nix, Sample dimensions influence strength and crystal plasticity. *Science*, 2004. **305**(5686): 986–989.

10 Other Testing Geometries and Conditions

Various loading geometries can be used for mechanical testing aimed at plasticity characterization. The simplest involve uniform stress states of uniaxial tension or compression, while the other common configuration is indentation, which creates complex and changing (2-D or 3-D) stress fields that are not amenable to simple analysis. These tests are covered in earlier chapters. However, other types of geometry can be employed, which may offer certain advantages. For example, bending or torsion of beams can be convenient experimentally and, while the associated stress fields are not uniform, they are relatively simple and may be suitable for analytical treatment. In fact, beam bending, in particular, offers potential for obtaining material properties via iterative FEM, in a similar way to indentation plastometry. Other geometries, such as those involving hollow tubes, may be relevant to particular types of application and expected (plastic) failure modes (such as buckling). There are also various tests involving temporal effects. Prolonged application of constant, uniform stress, leading to creep deformation, is covered in Chapter 5. However, again with a view to specific applications, the applied load may be cycled with a certain frequency, rather than being held constant or increased monotonically. While such (fatigue) testing is sometimes focussed on propagation of well-defined cracks, there is also interest in progressive damage that essentially arises from plastic deformation. Finally, some types of test are designed to create high strain rates, under which plasticity often takes place rather differently (because, as outlined in Chapter 3, the mechanisms involved exhibit a time dependence). This chapter covers all of these testing variants.

10.1 Bend and Torsion Testing

10.1.1 Mechanics of Beam Bending

Bending moments are produced by the application of transverse loads to beams. A simple example is the **cantilever beam**, widely encountered in balconies, aircraft wings, diving boards etc. A bending moment, M, acts on a section as a result of an applied transverse force. It is given by the product of the applied force and its distance from that section. It thus has units of N m.

The concept of the **curvature** of a beam is central here. Figure 10.1 shows that the axial strain, ε, is given by the ratio y/R, where y is the distance from the neutral axis and R is the radius of curvature. Equivalently, $1/R$ (the "**curvature**," κ) is equal to the

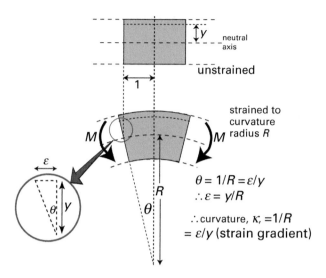

Fig. 10.1 Relation between the radius of curvature, R, the beam curvature, κ, and the strains within a beam subjected to a bending moment, M.

through-thickness gradient of axial strain. (Indeed, it is often preferable to think of the curvature as a strain gradient, rather than being concerned about the radius of curvature.) It follows that the axial stress at a distance y from the neutral axis of the beam is given by

$$\sigma = E\kappa y \tag{10.1}$$

where E is the Young's modulus.

At equilibrium, the applied moment, M, is balanced by the **internal moment** arising from the stresses generated within the beam. This is a summation of all of the moments acting on individual elements within the section. These are given by the force acting on the element (stress times area of element) multiplied by its distance from the neutral axis, y. This is illustrated by Fig. 10.2, which relates to a cantilever beam.

As shown in the figure, the (internal) moment can be written as

$$M = \int y(\sigma \, dA) = \int y(E\kappa y \, dA) = E\kappa \int y^2 dA \tag{10.2}$$

This can be presented more compactly by defining I (the **second moment of area**) as

$$I = \int_{y_{\text{inner}}}^{y_{\text{outer}}} y^2 dA \tag{10.3}$$

The units of I are m^4. The value of I is dependent solely on the beam sectional shape. Appendix 10.1 provides two examples of how a value of I can be obtained – for

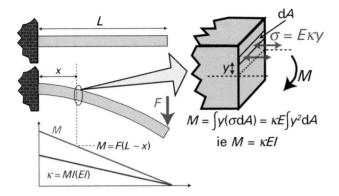

Fig. 10.2 Balancing the external and internal moments during the bending of a cantilever beam.

rectangular and circular section beams. Using this nomenclature, the moment can be written as

$$M = \kappa EI \tag{10.4}$$

The product EI is often called the "***beam stiffness***" and given the symbol Σ. It is the proportionality constant linking the "driving force" for bending (M) to the "outcome" (the curvature). This is analogous to the situation with simple axial extension, where the proportionality constant is E, linking the driving force (the applied stress) to the outcome (the strain). Of course, the moment may vary along the length of the beam (as it does for a cantilever), but these equations allow the resultant distribution of curvature (i.e. the shape of the beam), and the stress and strain distributions within it, to be calculated for any given set of applied forces.

Simulations of the distributions of curvature and stress that arise within cantilever beams or during three- or four-point bending, with a user-selected set of loads, is available in the "*Bending and Torsion of Beams*" package, which is one of many within the DoITPoMS set of educational resources (freely available under a creative commons arrangement). A screenshot from the page on which this particular simulation is available is shown in Fig. 10.3.

There is often interest in the ***deflections*** created during beam bending. For example, in some types of test it is this deflection that is measured in order to monitor the response of the sample. If the deformation is purely elastic, then these deflections can readily be related to the loading configuration and the elastic constants of the material. Examples of the calculations involved are shown in Appendix 10.2 (for an end-loaded cantilever and for symmetrical three-point and four-point bending).

10.1.2 Mechanics of Torsion

Torsion is the twisting of a beam under the action of a ***torque*** (twisting moment). It is systematically applied to screws, nuts, axles, drive shafts etc., and is also generated

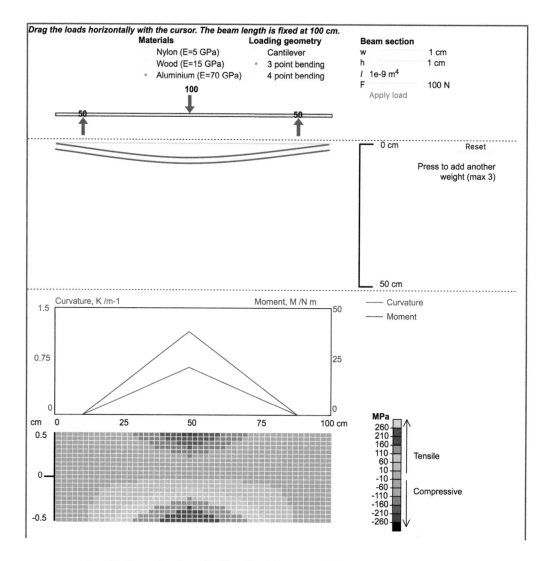

Fig. 10.3 Screenshot from the "Bending Moments and Beam Curvatures" page in the DoITPoMS TLP on *"Bending and Torsion of Beams,"* which is accessible at www.doitpoms.ac.uk/tlplib/beam_bending/bend_moments.php.

more randomly under service conditions in car bodies, boat hulls, aircraft fuselages, bridges, springs and many other structures and components. A torque, T, has the same units (N m) as a bending moment, M. Both are the product of a force and a distance. In the case of a torque, the force is tangential and the distance is the radial distance between this tangent and the axis of rotation.

As an example of this, torsion of a cylindrical bar is illustrated in Fig. 10.4. It can be seen that the shear strain in an element of the bar is given by

$$\gamma = r\frac{d\theta}{dL}$$

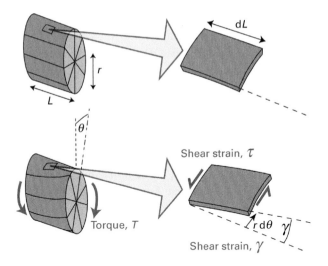

Fig. 10.4 Torsion of a cylindrical bar.

This equation applies both at the surface of the bar, as shown, and also for any other radial location, using the appropriate value of r. Clearly, the shear strain varies linearly with r, from zero at the center of the bar to a peak value at the free surface.

The shear stress, τ, at any radial location, is related to the shear strain by

$$\tau = G\gamma$$

where G is the shear modulus. It follows that

$$\tau = Gr\frac{d\theta}{dL}$$

The torque, T, can therefore be written as

$$T = \int_A dT = \int_A \tau r\, dA = \int_A Gr^2 \frac{d\theta}{dL} dA$$

As for the beam bending case, the geometrical integral is represented as a (polar) second moment of area:

$$I_P = \int_A r^2 dA \qquad (10.5)$$

For a solid cylinder of diameter w, this can be written as

$$I_P = \int_A r^2 dA = \int_0^{w/2} r^2 2\pi r\, dr = \left[\frac{\pi r^4}{2}\right]_0^{w/2} = \frac{\pi r^4}{32}$$

The torque is thus given by

$$T = GI_P \frac{d\theta}{dL} \qquad (10.6)$$

Comparing this equation with the corresponding one for beam bending (Eqn. (10.4)), it can be seen that the torsional analogue for the **curvature** of a bent beam is the **rate of twist** along the length of the bar. This can be measured experimentally, although not quite so easily as a curvature (because the macroscopic shape of the bar does not actually change – at least when it is straight: see Appendix 10.3 referred to below for an important example of a case when it is NOT straight).

An interesting example of torsion is provided by the deformation that takes place during the loading of *springs* (*torsional coils*). Of course, these have a wide range of engineering applications. They are normally made of (high yield stress) metals. When a spring is loaded (compressed or extended), the deformation experienced by the wire is one of pure torsion. This is described in Appendix 10.3. As outlined there, the configuration of a spring is an excellent one for creating (or accommodating) large deflections, while the (elastic) strains in the material remain low. This is convenient for measurement of elastic (and possibly plastic) properties, although of course it must be possible to form the sample into the shape of a coil.

10.1.3 Three-point and Four-point Bend Testing

Bending tests offer certain advantages over conventional uniaxial (tensile or compressive) configurations. Firstly, the sample can be in the form of a simple rod. If necessary, it could be in the form of a **non-prismatic beam** – i.e. one without a uniform sectional shape along its length, although this would be unusual and a rectangular or circular section rod is much more common. Secondly, there is no need for any gripping arrangement – it's only necessary to apply the (compressive) forces at well-defined locations. Thirdly, unlike the case of uniaxial loading, large deflections can be generated while the strains remain low, and the applied forces can also be relatively low.

To illustrate this, consider a square section rod, of side 5 mm and length 100 mm, of a material with a stiffness of 100 GPa and yield stress 100 MPa. If loaded in uniaxial tension up to the yield stress, the force required would be 2.5 kN and the extension would be 100 μm. Such displacements are often quite difficult to measure accurately, particularly since it would need to be separated from extensions due to various "bedding-down" effects and to elastic deformation in the loading train – see §5.2.2. If, on the other hand, the rod were to be loaded in three-point bending, then, at the "elastic limit" – i.e. when the beam is loaded so that the peak stress (in the center, at the free surfaces) reaches 100 MPa (i.e. the strain is 1 millistrain), and noting that the second moment of area, I, in this case (Eqn. (10.23)) is about 500 mm^4, Eqn. (10.4) can be used to show that the (peak) moment is about 20 N m. The necessary applied force is thus only about 400 N – i.e. about six times less than for tensile loading. Moreover, the central deflection, which can be obtained using Eqn. (10.31) in Appendix 10.2, would be about 200 μm. Not only is this larger than the extension during tensile testing, but it is easier to measure, since there are fewer complications relating to bedding-down, machine compliance etc. However, these are not necessarily entirely absent, since there could be penetration into the sample at the loading points and the loading train could have a compliance that is not entirely negligible compared to that of the sample.

There are thus certain advantages, compared with tensile testing, in using a bending configuration to measure elastic properties. In principle, this also applies to measurement of the "strength" of brittle materials. By noting the applied load at which the beam fractures, the corresponding peak stress in the sample can be obtained. Indeed, this methodology is employed quite widely. Unfortunately, there are difficulties. Of course, brittle fracture is sensitive to the presence of flaws, predominantly on free surfaces. Since the peak stress is created in only a small region during three-point bending, an unrealistically high apparent strength may be obtained. This issue is quite widely recognized and, for example, it is known that using a four-point bending configuration is likely to cause fracture when the peak stress is lower, because the volume of material in which that stress is being created is much greater than for three-point loading. However, since the main focus of this book is on plasticity, with little coverage of fracture, this issue is not taken any further here.

Concerning the characterization of plasticity, however, the details of what is taking place in the sample during bend testing are of interest, as are issues relating to design of the test. A bending rig is normally a fairly simple set-up that can be combined with a standard loading frame, with loading fixtures being commercially available. An example is shown in Fig. 10.5, where it can be seen how three- and four-point loading arrangements are created. It can also be seen that the standard arrangement for applying the load is via cylinders ("*rollers*"). This is experimentally convenient, allowing loads to be applied at well-defined positions along the beam (and also allowing the beam to rotate slightly while in contact with the rollers). However, it does mean that there is in principle a high contact stress (since the contact area is small). There may thus be a danger of some plastic deformation occurring there, probably in the sample, rather than the rollers. This may not create any serious

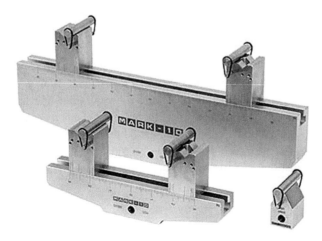

Fig. 10.5 Photo of typical fixtures for bend testing.

problems, although it will affect the load–displacement plot. In practice, however, it can often be avoided via suitable specification of the set-up dimensions.

In addition to such effects, clear detection of the point at which "controlled" plasticity starts to occur in the sample, by monitoring the load–displacement curve, is complicated by the fact that, in contrast to what happens in the tensile test, it will initially occur only in a very small volume. This volume will be larger during four-point bending than three-point, but it will still be small and in both cases the initial departure from linearity of the load–displacement plot may be hard to detect. This is explored in more detail in §10.1.4 below. (Of course, it should also be recognized that, while in principle generalized plasticity starts simultaneously throughout the sample in a tensile test, in practice the yield point may not be very well defined: this is explored in §4.3.1.)

The nature of the plastic strain fields during bend testing, and their effect on the load–displacement plot, has been explored using FEM modeling [1, 2]. Such work clearly reveals that, while the analytical treatments presented in §10.1.1 (and Appendix 10.2) are at least approximately correct, they are not necessarily very accurate, depending on the sample and loading system dimensions. For slender beams (having lengths much greater than their thickness), they should be a fair approximation, although in practice some samples may not satisfy this condition.

10.1.4 FEM of Four-point Bend Testing

As with other configurations, FEM simulation of bend testing can be useful in terms of understanding exactly what takes place within a sample, as opposed to the outcome of simple analytical treatments. A typical comparison between modeled and experimental load–displacement data [2] is shown in Fig. 10.6, for symmetrical four-point bending of an Al sample with the dimensions shown in the insert. It can be seen that, while the onset of plasticity ($\sigma_Y = 158$ MPa, with subsequent work hardening) is detectable, at least in the simulation, it is certainly not easy to observe or measure.

It may first be noted that the elastic part of this plot is approximately consistent with the analytical solution (Eqn. (10.35)). For example, using the I value for this beam of 3.33 mm^4, and a Young's modulus of 72 GPa, the predicted load at a displacement of 1 mm is just over 30 N, which is close to the value in Fig. 10.6. It follows that experimental measurement of this gradient should allow the value of E to be obtained with reasonable accuracy. The onset of plasticity, however, is much more problematic. The treatment in Appendix 10.2 can in principle be used to find the applied load at which the peak stress in the sample reaches the yield stress, when plasticity starts. This is done by noting that the peak strain, and hence the curvature (strain gradient), are given by

$$\varepsilon_{max} = \frac{\sigma_Y}{E}$$

$$\therefore \kappa = \frac{\varepsilon_{max}}{z_{max}} = \frac{\sigma_Y}{E z_{max}}$$

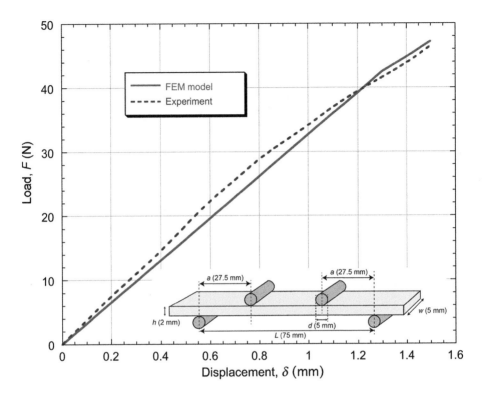

Fig. 10.6 Comparison [2] between measured and modeled data from four-point bend testing of an Al sample with the dimensions shown in the insert.

Using Eqn. (10.4) to relate the moment to the curvature:

$$M = \kappa EI = \frac{\sigma_Y}{z_{max}} I$$

Since the moment in the center is given by $Fa/2$ in this case, and using Eqn. (10.23) for the second moment of area, it follows that

$$F = \frac{2M}{a} = \frac{2\sigma_Y}{a\, z_{max}} I = \frac{\sigma_Y w h^3}{6a\, z_{max}}$$

Substituting $\sigma_Y = 158$ MPa, $w = 5$ mm, $h = 2$ mm, $a = 27.5$ mm and $z_{max} = 1$ mm, this gives the applied load as about 38 N. Actually, the FEM-predicted onset is at over 40 N, but in any event this onset is difficult to pick up from an experimental curve. As noted above, this plastic deformation only occurs in a small volume and the deviation from linearity is slight. This is true even for FEM-predicted curves, and is likely to be even more difficult to detect in experimental plots. Deducing a yield stress from a load–displacement plot, simply by using the analytical relationships, is therefore likely to be very inaccurate. Dong et al. also noted [2] that FEM-predicted distributions of stress differ significantly from those of the analytical model, particularly near the inner loading points, and that detectable effects can arise from changes in the radius of the contact rollers and in the sample surface roughness (affecting the coefficient of friction at the contact regions – the value used in obtaining Fig. 10.6 was 0.05, which is relatively low). It is in any event clear that, for accurate work, the process should be simulated using FEM models.

While obtaining plasticity characteristics (even just the yield stress) via the analytical equations is thus unlikely to be reliable, there is potential for using a similar approach to that of indentation plastometry – i.e. the plasticity characteristics (as captured in a constitutive law) can be obtained from tests of this type via inverse (iterative) simulation of the process. In fact, this concept can be applied to any test configuration. The target outcomes in this case are likely to be load–displacement plots, or possibly residual beam profiles. There may be scope for testing local regions in this way, although not on as fine a scale as for indentation. Of course, it could be argued that relatively small-scale tensile testing could be undertaken in a similar way, with no need for any inverse FEM. However, a bending configuration offers potential advantages. Machining a tensile sample on the scale of a few tens of mm is challenging, as is attempting to extract reliable strain data from a very short gauge length. A rectangular section beam, on the other hand, can readily be machined down to quite small scales and, since these large lateral displacements are easy to measure accurately, the sensitivity and reliability of the final outcome (stress–strain curve) are likely to be better than for a fine scale tensile test.

In fact, the four-point bend test configuration has already been used to investigate inhomogeneous samples. For example, Wang et al. [3] tested (relatively large scale) Al samples in the form of I-beams that contained welded regions. Modeling of the test with such samples requires much higher levels of complexity, concerning mesh characteristics as well as input data. Nevertheless, Wang at al. did show that a level of internal consistency could be obtained. For example, using the experimental (and simulated) stress–strain curves shown in Fig. 10.7 (for different locations relative to a weld, with quite substantial variations apparent), they obtained the levels of agreement shown in Fig. 10.8 between experimental and FEM-predicted load–displacement plots during four-point bending of samples with and without welds. With this test geometry, there is a tendency for buckling-type instabilities to arise – see §10.2 – and capturing of such effects required "imperfections" to be introduced into the model. In any event, there is clearly scope for FEM to be used to obtain information about the material response, even when both it and the test geometry are relatively complex.

10.1.5 Torsion Tests

Torsion testing is less popular than bend testing. There are several reasons for this. One is that gripping can present challenges. Another is that, apart from the use of a spring configuration (see Appendix 10.3), there are no well-defined displacements, although rotational motion of particular surface locations can be monitored. Thirdly, it cannot normally be used to monitor tensile fracture, since there are no well-defined planes across which tensile stresses operate.

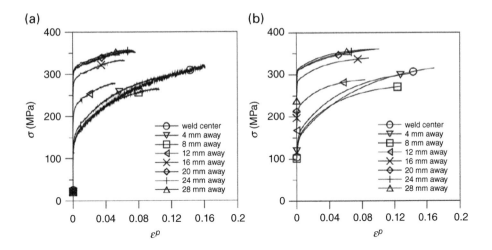

Fig. 10.7 Data for 6082-T6 Al [3] in the vicinity of a butt weld, showing (a) experimental true stress–true plastic strain curves and (b) best-fit simulated versions, obtained using a (three-parameter) constitutive law.

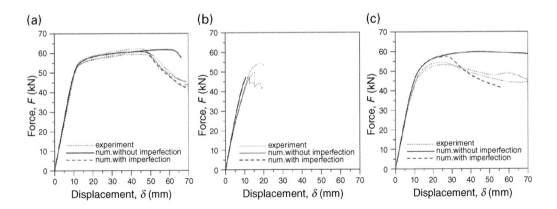

Fig. 10.8 Comparisons [3] between modeled and measured four-point bend load–displacement plots for samples containing (a) no weld, (b) a butt weld in the tension flange and (c) a butt weld in the compression flange.

This rather begs the question of whether torsion testing has any role (for plasticity characterization). It is in fact a very convenient approach to measuring the elastic properties of a metal if it is available in the form of a spring, to be subjected to axial extension or compression. In this form, gripping is not usually a problem and macroscopic displacements are generated, which are large and readily measured. The equations in §10.1.2 and Appendix 10.3 can be used to obtain the elastic properties from these. In fact, as for bend testing, while these displacements cannot be predicted analytically after the onset of plasticity, their measurement, in combination with the methodology of iterative FEM simulation, could be used to obtain complete stress–strain curves. On the other hand, the need to produce the sample in the form of a coil is a limitation and it would not be suitable for any study of anisotropy. In fact, not only would the need to form the sample into a coil be problematic in some cases, but also the forming operation might well affect its plasticity characteristics.

Nevertheless, torsion testing of metallic bars is carried out. The motivation for this is sometimes simply that it is close to service conditions for particular applications. However, this type of test does have some more basic potential [4, 5]. For example, high strain rates can be generated [6], although they are variable throughout the sample (except when the test is applied to thin-walled tubes, although in that case there might be a danger of a buckling type of local instability – see §10.2). In fact, it is not a common type of configuration for high strain rate testing, which is covered in §10.4, but it has been used to study strain rate effects. Furthermore, as for various other test geometries, inverse FEM can be used to obtain parameter values in constitutive laws and this has been done using torsion test data to obtain both quasi-static stress–strain curves [5] and strain rate sensitivity parameters [7, 8]. As with other types of high strain rate testing, changes in the thermal field during the test may be relevant [9, 10]. However, it offers no advantages over tensile testing in terms of sample preparation requirements and it is certainly not suitable for study of anisotropy, local property variations etc. In practice its usage is quite limited.

10.1.6 Combined Tension–Torsion Tests

While combined *tension–torsion* testing is not extensively used for plasticity studies, there are some situations in which it is useful. As an example of this, Fig. 10.9(a) shows a test configuration for the simultaneous application of tension and torsion to a (hoop-wound) composite sample [11]. For **highly anisotropic materials**, such as *uniaxial fiber composites*, the capacity to create customized stress states (relative to relevant directions or planes in the material) can be very helpful. For example, a range of combinations of transverse tension (σ_2) and shear, relative to the fiber axis, τ_{12}, can be applied. Figure 10.9(b) shows how testing in this configuration allows differentiation between two failure criteria. (The **Tsai–Hill criterion** incorporates potential for synergy between the failure modes of transverse tension and shear, while the **maximum stress criterion** does not.) In fact, further versatility could in principle be added by simultaneously applying internal pressure to the tubular sample, creating a hoop stress.

For metals, which in general do not exhibit anisotropy comparable to that of uniaxial fiber composites, there is less demand for such versatility, but a test configuration of this type might still be useful for some types of requirement, and analyses have been published [12, 13]. Furthermore, a particular metallic component might be in the form of a solid rod or tube, which will be subjected to combined tension/torsion/pressurization under service conditions. While FEM modeling could be used to simulate how the sample will respond, knowing its plasticity characteristics, it may still be reassuring to be able to carry out a test in which those conditions are created.

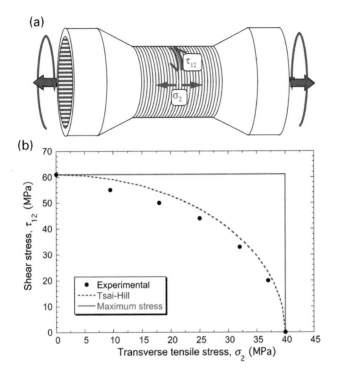

Fig. 10.9 Tension–torsion testing of hoop-wound composite tubes [11], showing (a) the test geometry and (b) a comparison between experimental data (for the failure of epoxy–65% glass composites) obtained in this way and predictions from two models.

10.2 Buckling Failure

10.2.1 Elastic (Euler) Buckling

The possibility of **buckling failure**, whether it involves some kind of fracture event or is a purely plastic collapse, often needs to be taken in to account when designing certain types of component or structure. The initial instability may be a purely elastic one, although it is likely to lead immediately to gross plastic deformation and/or fracture. Of course, this type of process is not solely dependent on material properties, but is also related to component dimensions and shape, as well as the loading configuration. Thus, while it's possible to identify a (yield) stress at which a rod under uniaxial compression will start to deform plastically, prediction of the applied stress at which it will buckle requires specification of its dimensions (length and sectional shape), and also of how it is constrained at the ends.

In fact, buckling of components under compressive loads has long been a topic of interest to structural engineers. A formula for the elastic buckling load of a column was derived by the Swiss mathematician Leonhard Euler in 1757. This is often written as

$$F = \frac{\pi^2 EI}{(kL)^2} \qquad (10.7)$$

where L is the length of the column and k is a dimensionless factor dependent on the **end constraint conditions**. For *"pin-jointed"* ends (that are free to rotate, but not to move laterally), k has a value of 1, whereas it is 0.5 for fixed ends (constrained from both rotating and moving laterally). Publications [14, 15] in which Euler buckling is explored theoretically date back to over a century ago and refinements continue to be added [16].

The product EI (the "beam stiffness" – see §10.1.1) is dependent on both the stiffness of the material and the sectional shape of the column. This buckling load can be expressed as a stress, often written as

$$\sigma_b = \frac{\pi^2 E}{s^2} \tag{10.8}$$

where s is the *"slenderness ratio,"* L/r, in which r is the *"radius of gyration"* ($= \sqrt{(I/A)}$, with A being the sectional area). This **buckling stress** thus falls off as $1/s^2$ – i.e. it is strongly dependent on the slenderness of the column. An idea of this sensitivity was, of course, obtained on a trial and error basis during millennia of experience with building of bridges, cathedrals etc. It is much more of a geometrical issue than one concerning material properties, although the Young's modulus is relevant and, for strongly anisotropic materials, such as wood, this becomes a more complex issue.

10.2.2 Plastic Buckling

It should, of course, be appreciated that plasticity can play an important part in buckling phenomena, and indeed that buckling can occur after the onset of global plasticity. The sample is effectively "less stiff" under these conditions, and hence more prone to buckling. There are well-defined tests in which samples are subjected to axial compression until some kind of buckling instability is stimulated and characterized. These can be fully solid samples, in which case it is common to observe some kind of global (long wavelength) instability, or they can be **hollow tubes** of some sort. In the latter case, depending on the ratio of diameter to wall thickness (and possibly on sample length), local plastic buckling in the walls is often stimulated, with much shorter wavelengths – often described as *"wrinkling."* Such tests are not normally undertaken in order to obtain basic plasticity characteristics, but they can be very useful in confirming exactly how failure occurs in particular cases. While it may be possible to predict this via numerical modeling, once the plasticity characteristics are known, in practice this can be a complex undertaking.

In contrast to elastic buckling, which is a well-defined, *catastrophic* effect, plastic buckling (of hollow structures) usually takes place *progressively* – perhaps over a (nominal) strain range of several %. This buckling can remain axisymmetric, although it's also possible for this symmetry to be lost as some kind of localized instability occurs. The stress–strain relationship observed during axial compression of a cylindrical tube does, of course, depend on that of the material, but the observed load–displacement plot will depart from this as buckling (wrinkling) sets in. This is illustrated schematically [17] in Fig. 10.10, where it can be seen that the shape of

10.2 Buckling Failure

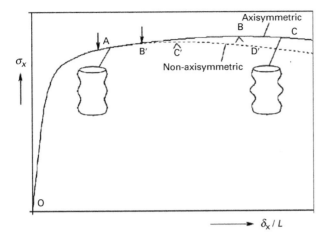

Fig. 10.10 Schematic plots [17] of (nominal) stress against strain during axial compression of a hollow circular-section cylinder, showing the onset of wrinkling (A), followed by axisymmetric collapse (BC) or non-axisymmetric collapse (B'C'D').

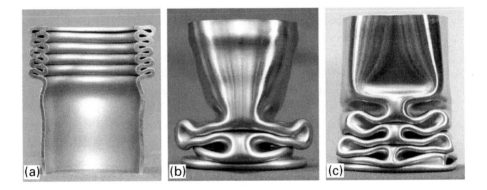

Fig. 10.11 Photos [17] of hollow tube samples after axial compression, showing (a) a carbon steel tube that has undergone axisymmetric concertina folding, (b) a stainless steel tube after non-axisymmetric (mode 2) folding and (c) another stainless steel tube after mode 3 folding.

the curve after the onset of instability will depend on whether axial symmetry is maintained. The fall in (nominal) stress as the test is continued is somewhat analogous to that seen in a tensile test after the onset of necking (§5.3). The visual appearance of samples after this type of test is illustrated [17] by those shown in Fig. 10.11.

Deformation (elastic and plastic) is uniform up to the point marked "A" in Fig. 10.10, when wrinkles start to form. If the deformation remains axisymmetric, then these grow in amplitude up to the point marked "B." Deformation then localizes, with the load dropping (BC). This localized deformation takes the form of one **axisymmetric lobe** growing until its folded walls come into contact. This can then occur with other lobes, giving **"concertina" folding**, as shown in Fig. 10.11(a). Alternatively, non-axisymmetric modes may develop, with two, three or more *circumferential waves* developing in the zone of localization, as shown in Fig. 10.11 (b) and (c), causing the load to start dropping earlier (B'C'D' in Fig. 10.10). This type of deformation tends to be favored with thinner walls (higher ratios of tube diameter, D, to wall thickness, t), although the stress–strain curve of the material is also relevant. Further details about observed and predicted behavior are available in the literature [18–20], although it's clear that these are difficult phenomena to capture reliably in (FEM) models. This is commonly the case with situations involving bifurcation phenomena. Nevertheless, this type of scenario is of considerable engineering significance, not least with regard to **energy absorption (*crashworthiness*)** during impact events [21–23] – when strain rate effects are also likely to be relevant.

10.2.3 Brazier Buckling of Thin-Walled Structures during Bending

A certain type of buckling instability is commonly observed when a ***thin-walled tube*** is subjected to a ***bending moment***. As the curvature increases, a point is reached when some kind of **kink** is created on the compressive side of the tube. This is easily observed by bending a drinking straw. It is a buckling instability, although the constraint conditions are rather different from those that apply during axial compression of a tube or cylinder. The phenomenon was first described by Brazier [24] in 1927 and his name is commonly associated with the effect.

If anything, experimental study and modeling of this effect is even more challenging than for buckling during axial compression of a tube, although, again, it is of considerable engineering significance. Its onset usually involves some kind of coupling between cross-section "*ovalization*," local **bifurcation buckling** and possibly the material plasticity characteristics. Brazier showed that progressive ovalization leads to an instability at a critical applied moment, usually termed the ***Brazier moment*** and given by [25]

$$M_{\text{Braz}} = \frac{2\sqrt{2}}{9}\left(\frac{E\pi rt^2}{\sqrt{1-v^2}}\right) = 0.987\left(\frac{E\pi rt^2}{\sqrt{1-v^2}}\right) \sim 1.035 E rt^2 \qquad (10.9)$$

where r, t are the radius and wall thickness of the tube and E, v are the Young's modulus and Poisson ratio (~0.3). The reliability of this has been confirmed in various subsequent treatments [26, 27], although it is found [25] that a bifurcation buckling event commonly takes place slightly before this ovalization limit (such that the constant in the equation is about 0.98). It may be noted that the material plasticity characteristics do not figure in these relationships, so in that respect this is being treated as an elastic buckling event (and indeed it is certainly much closer to that than the progressive wrinkling commonly observed during axial compression of tubes).

There are, however, a number of complications, some potentially related to the plasticity characteristics of the material. For example, strong work hardening may postpone full development of the buckle. There is also an issue related to the length of the tube. As might be expected, buckling becomes more difficult with shorter tubes, although the tube length, L, should be considered relative to its radius and wall

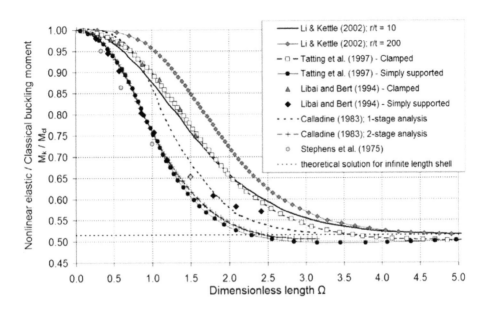

Fig. 10.12 Comparison [25] between several sources for the dependence of the buckling moment on the dimensionless length of isotropic elastic cylindrical tubes in bending.

thickness. However, this is again a complex issue from a modeling point of view. This is illustrated by Fig. 10.12, which shows predictions [25] from various models for the dependence of the buckling moment on the **dimensionless length** of the tube, defined as

$$\Omega = \frac{L}{r}\sqrt{\frac{t}{r}} \qquad (10.10)$$

It can be seen that, while the broad trend is the same for all of these models, they are certainly not in close agreement. This conveys an impression of the complexity of the modeling involved and, unfortunately, the difficulties of accurate experimental study are such as to make differentiation between models problematic. For the purposes of this book, it is sufficient simply to note that this type of failure can occur, but that it is not very sensitive to plasticity characteristics.

10.3 Cyclic Loading Tests

10.3.1 Background to Cyclic Loading and Fatigue Failure

There are many practical situations in which the load acting on a component or structure is neither constant nor monotonically increasing, but rather is cycled in some systematic way. Commonly, it is oscillated about some mean value, with a certain temporal frequency. The ratio of the amplitude of this oscillation to the mean value can vary over a large range, from representing a small perturbation to cases in which the variation is greater than the average, so that the **sign of the load reverses** on every cycle. It is, however, important to recognize that in some cases it may not actually be the load that is varying in a prescribed way, but rather that the sample or component is being subjected to imposed strains (or deflections) in a cyclic manner [28–30]. Temperature variations (differential thermal contraction) can create such **strain-controlled** conditions.

All such cases and tests are often referred to collectively as "fatigue" conditions, although the term does, of course, carry an implication that some kind of failure will occur after prolonged exposure. A distinction is sometimes drawn between "**high cycle fatigue**" (HCF) and "**low cycle fatigue**" (LCF). This categorization may be taken simply to relate to the number of cycles required for failure, with the HCF term being applied above a few thousand. However, it is sometimes taken to represent a slightly more fundamental distinction, with much more plastic deformation per cycle taking place during LCF, often due to larger cyclic variations in imposed strain or stress. It is occasionally stated that HCF relates to "elastic" conditions, although, if this were literally the case, then no change or damage would ever occur. In practical terms, while a complete spectrum of conditions is possible, there are certain types of application that would naturally fall into one or the other category. For example, the wings and fuselage of an aircraft are subjected to a relatively small number of loading cycles per flight (due to pressurization of the fuselage, take-off and landing, turbulence etc.), while each revolution of an internal combustion engine subjects components such as connecting rods to a cyclic load, with the number of cycles in a lifetime running into many millions. It's clear that these two examples would respectively be referred to as LCF and HCF conditions.

Since at least some fatigue failures occur as a result of progressive crack propagation, the basics of fracture mechanics and (sub-critical) crack growth now need to be summarized.

10.3.2 Fracture Mechanics and Fast Fracture

The key question, of course, is what exactly takes place during each cycle that leads to eventual failure. In some cases, this is the incremental advance of a well-defined crack. This is termed "**sub-critical crack growth**," to emphasize that the conditions do not satisfy the (Griffith) energy-based criterion for fast (unstable) crack growth [31]. This criterion is straightforward to derive. The energy released during crack advance comes from stored elastic strain in the surrounding material (plus any work done by the loading system). As a crack gets longer, the volume of stress-free material "shielded" by it from the applied stress increases – see Fig. 10.13.

The driving force for crack propagation therefore increases. The strain energy stored per unit volume in (elastically) stressed material is given by

$$U = \frac{1}{2}\sigma_0\varepsilon_0 = \frac{\sigma_0^2}{2E} \tag{10.11}$$

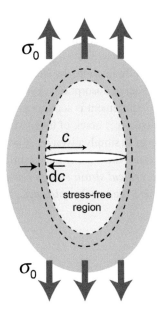

Fig. 10.13 Schematic depiction [11] of the stress-free region shielded by a crack from an applied load.

so the energy released when the crack extends (at both ends) by dc is the product of this expression and the increase in stress-free volume. The shape of the stress-free region is not well defined, and the stress was in any event not uniform within it before crack advance, but taking the relieved area to be twice that of the circle having the crack as diameter gives a fair approximation. Thus, for a plate of thickness t, the energy released during incremental crack advance is given by

$$dW = \frac{\sigma_0^2}{2E} 2(2\pi ct\, dc) = \frac{2\sigma_0^2 \pi ct\, dc}{E} \quad (10.12)$$

A central concept in fracture mechanics is that of stored elastic strain energy being released as the crack advances. The **strain energy release rate** (crack driving force) is usually given the symbol G (not to be confused with shear modulus or Gibbs free energy). It is a "rate" with respect to the creation of new crack area (and so has units of J m^{-2}) and does not relate to time in any way. It follows that

$$G = \frac{dW}{\text{new crack area}} = \frac{2\sigma_0^2 \pi ct\, dc/E}{2t\, dc} = \pi\left(\frac{\sigma_0^2 c}{E}\right) \quad (10.13)$$

The value of the constant (π in this case) is not well defined. It depends on specimen geometry, crack shape/orientation and loading conditions. In any event, the approximation used for the stress-free volume is simplistic. However, the dependence of G on ($\sigma^2 c/E$) is more general and has important consequences, particularly in terms of the linear dependence on crack length. It may also be noted at this point that the "crack area" concept in this treatment refers to a "***projected area***" – no attempt is made to monitor the actual crack area, which may, of course, be greater than the projected area if the crack is repeatedly undergoing minor changes of direction – being deflected, for example, by microstructural features.

In order for crack propagation to be energetically favored, the strain energy release rate must be greater than or equal to the rate of energy absorption, expressed as energy per unit (projected) area of crack. This energy requirement is sometimes known as the **Griffith criterion**. Focussing now on an edge (surface) crack of length c, propagating inwards, for a brittle material this *fracture energy* is simply given by 2γ (where γ is the *surface energy*, with the factor of 2 arising because there are two new surfaces created when a crack forms). It can be considered as a **critical strain energy release rate**, G_c. It is a material property. It is sometimes termed the **crack resistance**. The *fracture strength* can thus be expressed as follows:

$$G \geq G_c = 2\gamma$$

$$\therefore \pi \left(\frac{\sigma_0^2 c}{E} \right) \geq 2\gamma \tag{10.14}$$

$$\therefore \sigma_* = \left(\frac{2\gamma E}{\pi c} \right)^{1/2}$$

This equation can be used to predict the stress at which fracture will occur, for a component containing a crack of known size. However, it relates only to materials for which the fracture energy is given by 2γ – i.e. for which the energy absorbed during crack propagation is only that needed to create the new surface area. It may be noted at this point that the magnitude of 2γ does not vary very much between different materials and is always relatively low – usually $<\sim 10$ J m^{-2}. These are regarded as "ideally brittle" materials. Some materials, including many glasses, do behave in at least approximately this way, but most materials do not.

It has long been clear that the simple Griffith condition (Eqn. (10.14)) applies only to brittle materials. For metals, a fracture event, often preceded by extensive plastic flow throughout the sample, and then by localized necking, is usually very different from that in brittle materials. Apart from the possibility of energy being (permanently) absorbed by uniform plasticity before any crack growth occurs, the crack propagation process itself requires much more energy than in brittle materials. There tends to be a zone of plasticity ahead of the crack tip, irrespective of whether there has been much plastic deformation prior to the onset of fracture. Figure 10.14 shows how this plasticity raises the radius of curvature, r, at the crack tip and reduces the stress concentration effect. (The peak stress, indicated as σ_Y in Fig. 10.14(b), is nominally the yield stress, although it will be higher than this if work hardening occurs.)

The work done during plastic deformation must be supplied by the **crack driving force** (release of elastic strain energy). It might be imagined that a simple modification could be made to Eqn. (10.14), replacing 2γ by an alternative version of G_c, which takes account of these extra energy requirements (and, indeed, G_c values for tough materials such as metals are commonly greater than 2γ by several orders of magnitude). This apparently allows the Griffith criterion to be applied to ductile materials, but there are difficulties. It suggests that ductile materials should be as sensitive to the

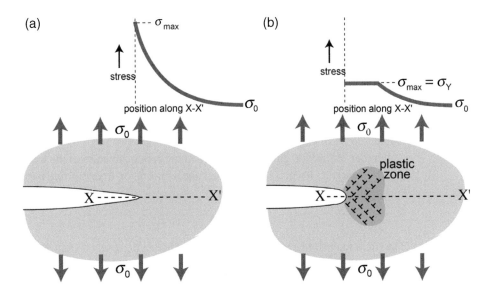

Fig. 10.14 Crack tip shapes and stress distributions [11] for (a) brittle and (b) ductile materials.

presence of flaws as brittle materials, although they would have higher fracture stresses for a given crack size. In fact, failure stresses are not dramatically or systematically higher for ductile materials, although they certainly fracture less readily and require much more energy input. Furthermore, ductile materials show little or no sensitivity to initial flaw size. Very ductile materials often fail by **ductile rupture** (progressive necking down to a point), with little or no crack propagation as such. The failure stress in such cases can only be predicted by analysis of the plastic flow and Eqn. (10.14), with 2γ replaced by a much larger value of G_c, is irrelevant. However, such highly ductile materials are too soft to be useful for most purposes. Failure of engineering metals does commonly involve fracture, often after extensive plastic flow and some necking have occurred, as described in earlier chapters. Clearly, use of Eqn. (10.14) requires care in such cases. The appropriate fracture energy value would relate to material that had already undergone extensive plastic deformation and the stress would be the true value acting in the neck. Neither of these is well-defined. (In practice, it is common to simply assume that final fracture will occur when the plastic strain in the neck reaches a critical value, and indeed this is often a plausible assumption, although the *critical strain* certainly cannot be defined universally.)

10.3.3 Sub-Critical Crack Growth

When the rate of energy release (driving force) during crack propagation is lower than the critical value, spontaneous fast fracture does not occur. Under some circumstances, however, an existing crack may advance **progressively** under this driving force. Since crack growth increases the driving force (for the same applied load), this process is likely to lead to an accelerating rate of damage, culminating in conditions for fast fracture being satisfied. There are two common situations in which such sub-critical crack growth tends to occur. Firstly, if the applied load is fluctuating in some way, local conditions at the crack tip may be such that a small advance occurs during each cycle, moving through a region in which high plastic strains have been created. Secondly, the penetration of a corrosive fluid to the crack tip region may lower the local toughness and allow crack advance at a rate determined by the fluid penetration kinetics or chemical interaction effects. In both cases, as indeed with fast fracture, the presence or absence of an initial flaw, which allows the process to initiate, is likely to be of considerable importance.

In order to link the global energy-based concept of fracture to the local conditions at the crack tip, Irwin [32] proposed the concept of a **stress intensity factor**, K, such that

$$K = \sigma_0 \sqrt{\pi c} \qquad (10.15)$$

The stress intensity factor, which has units of MPa √m, scales with the level of stress at the crack tip (although it does not allow the absolute value to be established). Fracture is expected when K reaches a critical value, K_c, the **critical stress intensity factor**, which is often termed the **fracture toughness**.

There are clearly parallels between K reaching a critical value, K_c, and G reaching a critical value of G_c. The magnitude of K can be considered to represent the crack driving force, analogous to G. Consider again the Griffith energy criterion:

$$G \geq G_c, \text{i.e. } \pi \left(\frac{\sigma_0^2 c}{E} \right) \geq G_c$$

$$\therefore \sigma_0 \sqrt{\pi c} \geq \sqrt{E G_c} \qquad (10.16)$$

It can be seen that this is similar in form to Eqn. (10.15). It follows that

$$K = \sqrt{EG} \geq \text{ and } K_c = \sqrt{EG_c} \qquad (10.17)$$

While it is reassuring to be able to treat fracture from both stress and energy viewpoints, it is not immediately apparent what advantages are conferred by using a stress intensity criterion, rather than an energy-based one. However, in practice it is possible to establish stress intensity factors, and corresponding fracture toughness values, for various loading and specimen geometries, whereas this is not really possible with an energy-based approach. The 3-D stress state at the crack tip, and hence the size and shape of the plastic zone, can be affected by specimen thickness and width. In fact, an important point concerning experimental measurement of G_c and K_c is that it should normally be done under conditions of **plane strain**, which effectively requires the sample to be relatively wide and also deep (so that the crack is not close to the back surface). The size of the plastic zone at the crack tip can become larger if these conditions are not maintained, raising the measured toughness.

A further issue concerning use of the stress intensity factor is that rates of **sub-critical crack growth** can often be predicted from K, since it is directly related to conditions at the crack tip, whereas the strain energy release rate is a more global parameter. As an example of this, it can be shown that the size of the *plastic zone* ahead of the crack tip is related to the yield stress of the material by

$$r_Y \sim \frac{1}{2\pi}\left(\frac{K}{\sigma_Y}\right)^2 \tag{10.18}$$

Similarly, the crack opening displacement, δ, can be expressed as

$$\delta \sim \left(\frac{K^2}{E\,\sigma_Y}\right) \tag{10.19}$$

Such parameters are useful when considering how energy-absorbing processes might be stimulated, since they allow the scale of features of the crack tip to be related to the scale of the microstructure.

Analysis of fatigue is often carried out in terms of the difference in stress intensity factor between the maximum and minimum applied load (ΔK). This is because, while the maximum value, K_{max}, dictates when fast fracture will occur, the cyclic dissipation of energy is dependent on ΔK. It is, however, common to also quote the *stress ratio*, $R\ (= K_{min}/K_{max})$, which enables the magnitudes of the K values to be established for a given ΔK. The resistance of a material to crack extension is given in terms of the crack growth rate per loading cycle (dc/dN). At intermediate ΔK, the crack growth rate usually conforms to the *Paris–Erdogan relation* [33]:

$$\frac{dc}{dN} \sim \beta \Delta K^n \tag{10.20}$$

where β is a constant. Hence, a plot of crack growth rate (m/cycle) against ΔK, with log scales, gives a straight line in the *Paris regime*, with a gradient equal to n. At low stress intensities, there is a *threshold*, ΔK_{th}, below which no crack growth occurs. The crack growth rate usually accelerates as the level for fast fracture, K_c, is approached.

It should be emphasized that, while an overview has been presented here concerning fact fracture, and sub-critical crack growth under fatigue loading, the focus of this book is on plasticity, with no detailed coverage of fracture or crack propagation. The section below is thus oriented towards cyclic loading under conditions in which there are no pre-existing or well-defined cracks, but rather there is progressive damage – often due to the local creation of high plastic strains (and possibly the formation of multiple cracks). Such testing is in many cases of direct relevance to industrial applications. It should also be appreciated that, as with most types of cyclic loading, it is mainly in the near-surface regions that the important phenomena take place.

10.3.4 Stress–Life (S–N) Fatigue Testing

An alternative way of presenting fatigue data is in the form of S/N_f curves, showing the number of cycles to failure (N_f) as a function of the stress amplitude (S). Many materials exhibit rapid failure (low N_f) when the stress amplitude is high, a central portion of decreasing S with rising N_f, possibly corresponding to the Paris regime, and a *fatigue limit*, which is a stress amplitude below which failure does not occur, even after large numbers of cycles. This can be interpreted as corresponding to a stress intensity factor below the threshold, ΔK_{th}. However, it should be noted that one advantage of the S–N formulation is that it does not involve any assessment of crack lengths, and so is better suited to cases in which there is no propagation of a single, well-defined crack.

Of course, it should be appreciated that, experimentally, the tests are often simply continued until the sample fails (so that N_f, for a given S, is the sole outcome of the test), with no information supplied about issues such as whether there was a single dominant crack from an early stage. Also, at least for most S–N testing, it is assumed that the stress is simply cycled from zero to the value concerned, so that the concept of a stress ratio does not arise. Furthermore, while a frequency of cycling must be chosen, it is rarely considered to be a variable that might affect the outcome. It should perhaps be noted that a possible role of frequency concerns heat generation (associated with plastic deformation). If this is significant, then the cycling frequency could have an effect on any associated rise in local temperature. This is sometimes considered to be an issue for polymers (and other low conductivity materials), but for metals it is commonly assumed that the temperature remains uniform and constant.

For samples in which there is no "pre-crack" – and of course industrial components would rarely have anything of that sort – it is clear that **crack initiation** is a key step towards failure, whether or not a single crack becomes dominant. This commonly occurs as a result of **excessive plasticity** in the near-surface region and there is frequent reference [34–37] in the literature to "*intrusion–extrusion*" (plasticity) mechanisms for creation of surface defects that develop into cracks. Of course, softer materials tend to be more susceptible to this. Also of potential relevance is the presence of **residual stresses** in near-surface regions. The beneficial effect of compressive residual stresses, which are commonly created by "*shot peening*" or "*laser peening*," is very well documented [38–42]. As an example [43] of this, Fig. 10.15 shows (a) a typical sub-surface distribution of residual stress after shot peening and (b) the measured effect of this treatment on the fatigue life of an Al alloy, for several different applied stress amplitudes. These residual stresses tend to become relaxed after extended cyclic testing, but it's nevertheless clear that the (beneficial) effects in retarding failure can be substantial.

Of course, microstructural issues are often very important, particularly for ferrous alloys (in which there is considerable scope for control over near-surface properties). There are several options for surface hardening, such as **carburizing** or **nitriding**, and these can be very effective [44–48] in retarding the initiation of cracking under fatigue loading. There have also been many studies [49–51] in which the focus is on potentially deleterious microstructural features, such as non-metallic **inclusions**, which might nucleate cracks. There is certainly evidence that the presence of such inclusions can reduce the fatigue life, particularly if they are relatively large. For example, the compilation of data [49] shown in Fig. 10.16 provides evidence for this.

As with many other types of material property, there have been attempts to fit fatigue life data to empirical laws. One of the simplest of these is the **Basquin law**, which is usually written [52]

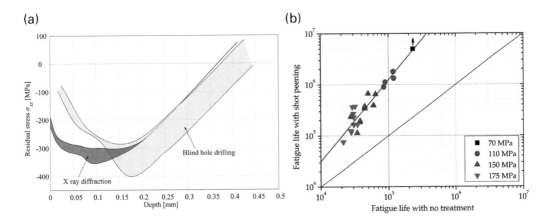

Fig. 10.15 The effect [43] of shot peening-induced compressive residual stresses on the fatigue life of Al-7075 alloy, showing (a) the measured sub-surface distribution of (equal biaxial) residual stress and (b) the effect of this treatment on the fatigue life, for several different applied stress amplitudes.

$$S = a N_f^b \tag{10.21}$$

where b is the Basquin exponent, which is normally negative and has a magnitude of less than one. Experimental data typically do exhibit behavior of approximately this type – i.e. the N_f value increases as the stress level is dropped. In practice, however, there is often a *"fatigue limit"* – i.e. a stress level below which failure never occurs. A modified version of this law was therefore proposed [52] by Stroymeyer, with the form

$$S = a N_f^b + S_L \tag{10.22}$$

where S_L is this limiting stress. Of course, this is entirely empirical, with little in the way of mechanistic justification. There have, however, been various attempts [53–55] to introduce modified expressions and to rationalize the values of constants in them, using parameters such as the hardness of the material and the size of inclusions present in it. As an example of this, Fig. 10.17 compares S–N data [55] from six different steels with ranges predicted on this basis by two "models" (although these don't actually incorporate fatigue limits). The real predictive capability of such formulations is in fact quite limited, since they take no account of details of the test conditions, surface finish, residual stresses, cycle frequency and various other parameters that could be relevant. Nevertheless, they are potentially of value in a semi-quantitative way for identifying trends.

10.4   Testing of the Strain Rate Dependence of Plasticity

A brief outline is supplied in §3.2.2 concerning constitutive laws that describe the dependence of plastic deformation characteristics on the imposed strain rate. Most

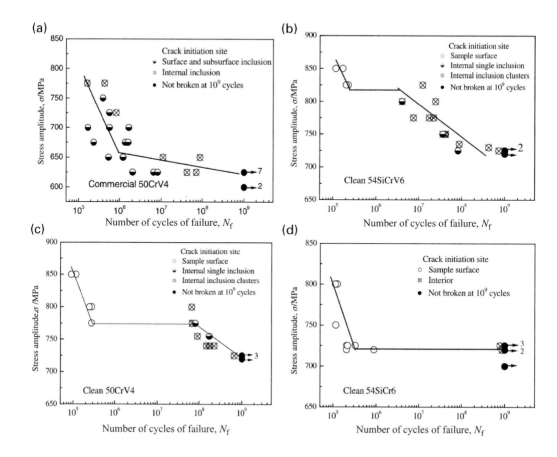

Fig. 10.16 S–N curves [49] for four steels, with indications about crack initiation sites for each test. These were inclusions in many cases, except for steel (d), which was the only one to contain no large inclusions. Strengths were lower for steel (a), which was the only one not to be "cleaned," and consequently contained many large inclusions.

metals effectively become harder when deformed at high strain rates, although this tends to become a significant effect only when the rate becomes quite high ($>\sim 10^3$ s^{-1}). An explanation for this in terms of the mechanisms of plastic deformation is provided in Chapter 4, particularly in §4.4. The most common constitutive law governing this effect is that of Johnson and Cook, which is presented as Eqn. (3.15). There is essentially just a single dimensionless parameter, C, that characterizes the sensitivity to strain rate, although another is required that relates to the temperature dependence of the quasi-static plasticity (since temperatures commonly rise during high strain rate deformation). Alternative formulations have in fact been proposed, but these do not introduce radically different concepts and in the sections below the focus is on experimental procedures that can be used to evaluate the Johnson–Cook parameter.

10.4 Strain Rate Dependence of Plasticity

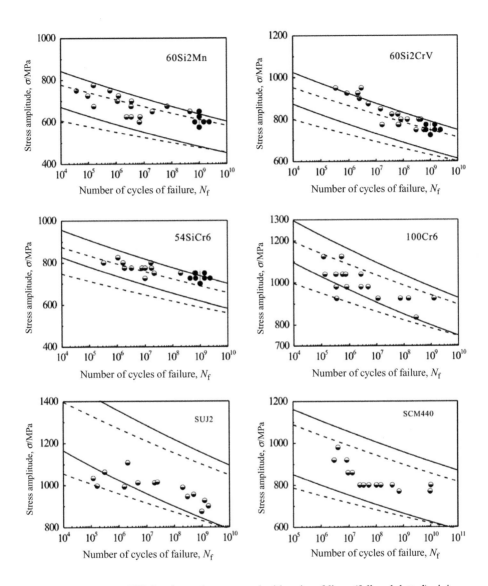

Fig. 10.17 S–N curves [55] for six steels, compared with pairs of lines (full and dotted) giving bounds for the relationship, based on two different empirical formulations that incorporate measurable parameters such as hardness and average inclusion size.

10.4.1 High Strain Rate Tensile Testing

Conventional mechanical testing procedures have certain limitations for this purpose and the maximum strain rates achievable in a controlled way during uniaxial tensile or compression testing are below those that are often of interest. There are several issues [56–58], but the main problem is that, in order to impose high strain rates on a bulk sample, the cross-head must move (and then stop!) very quickly. This issue is mentioned in §5.4.3. It's difficult to create the necessary accelerations and

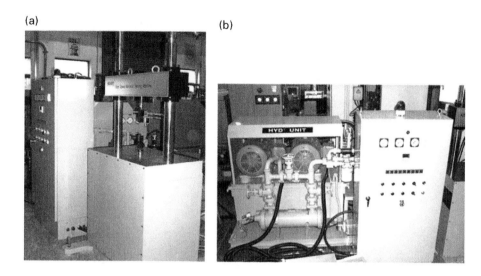

Fig. 10.18 Photos [59] of a high speed tensile testing machine, showing (a) the loading frame and (b) the servo-hydraulic unit.

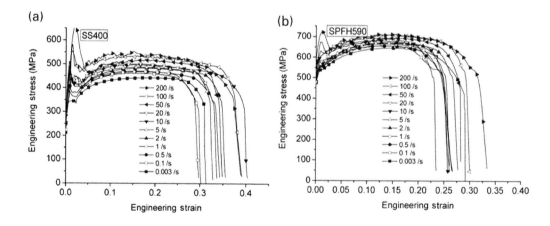

Fig. 10.19 Engineering (nominal) stress–strain plots [59] obtained during high speed tensile testing of (a) SS400 and (b) SPFH590 steels.

decelerations, particularly since cross-heads are often relatively massive. As an example, for a sample 10 mm in length, the cross-head must be moving at about 10 m s^{-1} in order to create a strain rate of 10^3 s^{-1}.

Speeds of this order can be generated, for example using hydraulic pressure with accumulators, as shown [59] in Fig. 10.18, but it is certainly not easy. Furthermore, it's not possible to accelerate the cross-head instantaneously to the required velocity, so a special jig is needed, such that the sample is only gripped once this velocity has been reached [59]. In practice, even with such arrangements, maximum strain rates are usually of the order of a few hundred s^{-1}. Strain rate effects on stress–strain curves can be detected in that range, as shown [59] in Fig. 10.19. In such plots, necking starts at the peak and little significance can be attached to the post-necking regime (§5.3.3 and §5.3.4), but detectable rises in flow stress are apparent over this range of strain rate. However, much higher rates than this are generated during various impact events of industrial and commercial interest, so other types of test are certainly needed.

10.4.2 Hopkinson Bar and Taylor Tests

There are a number of tests designed to create genuinely high strain rates (typically in the range $\sim 10^3$–10^6 s^{-1}) in a controlled manner, so that a stress–strain curve, or at least the change in some average flow stress value (relative to quasi-static conditions) can be obtained. As outlined above, creating such conditions while the sample is being subjected to a uniform tensile stress is virtually impossible, so the emphasis has been on compressive loading or on creating a more complex, non-uniform stress field, such as those that arise during bending or torsion – see §10.1.1 and §10.1.2. One possible arrangement is to uniaxially compress a sample using a falling weight or other impact arrangement, using the measured deceleration of the impacting body to monitor the response of the sample. However, as outlined in Chapter 6, complications can arise from frictional effects, and from the changing geometry, if relatively large strains are developed. In fact, the main difficulty with virtually all such configurations arises from the danger of the stress and strain fields changing qualitatively during the test, complicating the interpretation of output data. It's also important to distinguish tests designed to investigate plasticity from those aimed at obtaining fracture toughness information (which are sometimes based on very similar test geometries).

Originating in the 1940s and 1950s, there are several configurations that are usually grouped together as *"Hopkinson Bar"* tests, although the terminology can be somewhat confusing. The key elements of one of the central designs, usually termed the *"Split Hopkinson Pressure Bar"* (SHPB) are shown [60] in Fig. 10.20. The sample is a short solid cylinder, which is subjected to axial compression. (The term "split" is used because the sample sits in a loading train composed of other bars.) When a pressure pulse (created via a gas gun) reaches the sample, part of it is reflected and part transmitted. Incident and reflected pulses, and also the transmitted pulse, are picked up by strain gauges on the incident bar and the transmitted bar. These bars are made of a

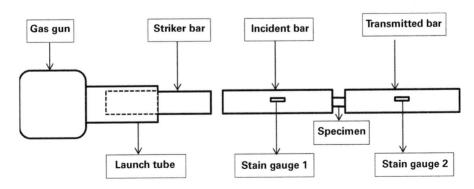

Fig. 10.20 Schematic representation [60] of a Split Hopkinson Pressure Bar test set-up.

hard material (such as a maraging steel) that remains elastic. The signals are analyzed [60, 61] so as to obtain the stress, strain and strain rate generated in the sample. The strains are relatively small, such that the geometry of the loading does not change much during the test. As might be expected, there are a number of potential sources of error and there have been many studies [60, 62–64] of these and of measures that might reduce their significance.

There are several variants of the SHPB test. Prominent among these is the *"Torsional Split Hopkinson Bar"* [65–67] configuration (TSHB). There are again bars on both sides of the sample, which now transmit torque to it, rather than axial compression. There are several options regarding the mechanism by which a (large) torque is applied to the sample, but a common one is to use a lathe bed, setting the complete assembly rotating and then rapidly arresting the rotational motion of the bar on one side of the sample. There are also designs involving flywheels and electromagnetically induced forces.

If the sample is in the form of a solid cylinder, then the stress field varies with radial location, as described in §10.1.2, but in practice it is common to use hollow tubular samples. This creates what is often described as a state of pure shear in the sample. In principle, this corresponds to a biaxial stress state in which the two principal stresses are of equal magnitude, but opposite sign – i.e. such that the Mohr's circle is centered on the origin (see §2.2.3). If the sample is relatively thin-walled, then the stress state is similar throughout it. This is illustrated schematically in Fig. 10.21. The output data are again obtained via the readings of strain gauges on the two rods adjacent to the sample (which are assumed to remain elastic).

It may be helpful at this point to note how **strain gauge readings** are converted to **principal strains**. For the SHPB configuration, it may be sufficient to just have single gauges on the two rods, aligned along the axis of the system (expected to be a principal direction). For the TSHB test, however, the rods are being subjected to shear. For such cases, and indeed for any general case, **strain gauge rosettes** are

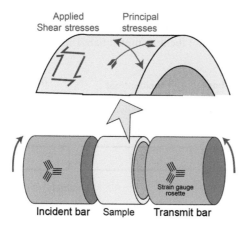

Fig. 10.21 Stress state created in a Torsional Split Hopkinson Bar (TSHB) test.

commonly used, rather than single gauges. The principle strains, and their orientation, can then be obtained from the output of the three gauges, as described in Appendix 10.4. Of course, for both SHPB and TSHB testing, these strain gauge signals are highly transient, so that relatively fast data logging is often required in order to analyze the response of the sample.

Mention should also be made of the ***Taylor cylinder test*** [68–71]. This involves similar conditions to the SHPB test – i.e. in principle it creates uniaxial compression in a solid cylindrical rod. However, this is done without using adjacent bars to either apply the load or to monitor the response of the sample. The sample is usually struck by a disk or bar of some type (made of a hard material), with the motion commonly being created using a gas gun. Furthermore, the outcome of the test is the deformed shape of the sample, with measurements taken during the experiment usually being limited to monitoring of the impact velocity. Strain rate sensitivity information is commonly obtained via inverse FEM modeling of the deformation, using a constitutive law and seeking optimal agreement between measured and modeled sample shape.

There have in any event been many publications reporting on outcomes of testing of Hopkinson and Taylor bars, often involving evaluation of the Johnson–Cook strain rate sensitivity parameter (C). Values of C obtained in this way include 0.001 for a 1000 series Al alloy [72], 0.023 for a 7000 series Al alloy [73], 0.039 for an AISI-1018 low carbon steel and 0.011 for an AISI-4340 low alloy steel [74], 0.048 for an AISI-1018 low carbon steel [75] and 0.009 for an ultra-fine grained copper [76]. It's difficult to rationalize such outcomes in any way and even this small cross-section of results indicates that they are not always entirely consistent. Nevertheless, it's clear that some metals exhibit considerably greater strain rate sensitivity than others. It is difficult to explain such variations on the basis of microstructure and deformation mechanisms, although it is plausible that softer material may exhibit more strain rate hardening.

These tests are quite widely applied, but they do suffer from certain disadvantages. Firstly, the constraint on the plastic deformation is often rather ill-defined. This can introduce errors into the modeling, due to uncertainty surrounding the exact boundary conditions that should be used. Secondly, various pressure/shock waves are created and propagated, which can complicate analysis of the output data. Thirdly, the equipment required is often rather cumbersome and poorly standardized, with relatively large amounts of energy involved (and potential for ***safety issues*** to arise). In view of this, there is interest in ballistic indentation, in which only a small fraction of the sample is deformed, conditions are better defined and energy levels tend to be relatively low. This is described in the next section.

10.4.3 Ballistic Indentation Testing

As is emphasized a number of times in this book, the inverse FEM method is a potentially powerful approach to obtaining plasticity parameter values. Its application to ballistic indentation for characterization of the strain rate sensitivity is particularly attractive, since a wide range of local strain rates can be generated and there are few uncertainties in the formulation of the model (as a *free body problem*). Furthermore, the requirements in terms of sample dimensions and preparation procedures are less demanding. It is also potentially easier than the tests described in §10.4.2 in terms of experimental implementation (provided that suitable arrangements for monitoring of experimental outcomes can be employed).

One of the challenges in implementing iterative FEM procedures concerns the methodology for converging on an optimized set of parameter values. With multiple parameters, this can be relatively complex and slow. However, in the Johnson–Cook constitutive law (see Eqn. (3.15) in §3.2.2) only one parameter (C) needs to be evaluated, provided that the quasi-static plasticity parameters (or a set of data pairs) have already been obtained. This simplifies the convergence procedure considerably. As an example of this, some data are presented from an experimental investigation [77] in which hard spheres were projected at annealed and work-hardened copper samples. Monitoring of the outcome was done via both high speed photography and measurement of the residual indent profile. The experimental arrangement is shown in Fig. 10.22.

A potential difficulty in modeling of high strain rate conditions (either during a test or in an industrial application) arises from the fact that it is common for relatively high strains to be developed (perhaps hundreds of %). Before strain rate effects can be quantified, the corresponding quasi-static stress–strain curve is required and it is far from simple to obtain this up to such strain levels using conventional (uniaxial) tests. Potentially, there is scope for obtaining such information via indentation plastometry (Chapter 8), which allows higher plastic strains to be generated in a controlled way

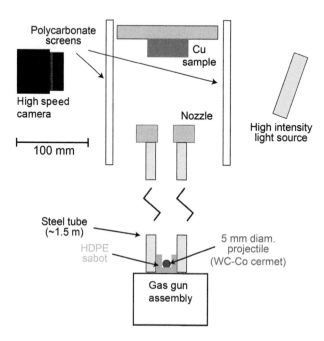

Fig. 10.22 Schematic depiction [77] of a set-up for ballistic indentation.

(although obtaining good sensitivity to the response at very high strains is not easy). In the study concerned [77], stress–strain relationships up to 300% strain were obtained partly via compressive testing of samples subjected to large prior plastic strains via swaging.

Some information about the conditions during impact can be obtained from the fields of plastic strain, strain rate, von Mises stress and temperature shown in Fig. 10.23 (for

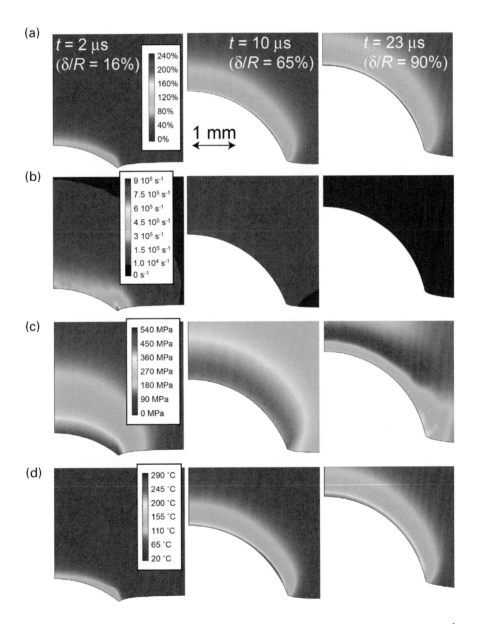

Fig. 10.23 Predicted [77] FEM fields for annealed Cu, with a projectile velocity of 200 m s^{-1} (and an optimized C value of 0.03), showing (a) equivalent plastic strain, (b) strain rate, (c) deviatoric stress and (d) temperature, at three different times after initial impact.

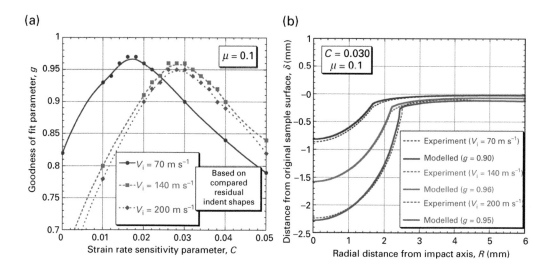

Fig. 10.24 Data [77] from residual indent profiles after impact with different velocities, for samples of annealed Cu, showing (a) dependence of the goodness of fit parameter on the value of C and (b) comparisons between measured and modeled data, for the best-fit value of C.

an impact velocity of 200 m s^{-1}). Among the points to be noted there are that strains can indeed reach hundreds of %, strain rates can be the order of 10^5–10^6 s^{-1} and temperatures can rise significantly. For the annealed and work-hardened Cu used in the work, the optimized values of C were respectively 0.030 and 0.016, with good consistency between use of measured displacement histories during impact (via high speed photography) and use of the residual indent profile. Based on residual indent profiles only, Fig. 10.24 shows goodness of fit as a function of C value, together with a comparison between measured and modeled profiles for the best-fit value, for the annealed Cu. Figure 10.25 shows corresponding data for the work-hardened (as-received) material.

It can be seen that these data show good consistency and sensitivity. In fact, since the sensitivity is good for use of residual indent profile data, there is real potential for the creation of an experimental set-up and methodology for obtaining strain rate sensitivity information (without requiring high speed photography, and possibly without the need for a conventional gas gun as such). A set-up similar to that for basic indentation plastometry (§8.4.6) could be used, with provision for high speed impact of a (reusable) indenter in a housing. In fact, if a sphere of 2 mm diameter were used, rather than the 5 mm diameter projectile of the above work, then similar strain rates could be created with lower impact velocities, while still allowing suitably accurate profile measurement. Possibly such incident velocities could be created using a spring-based or electromagnetic system. It might be necessary to assume a value for m (in Eqn. 3.15), which controls the influence of temperature, and also for the coefficient of friction, μ, but the overall sensitivity of the outcome to these values is relatively low.

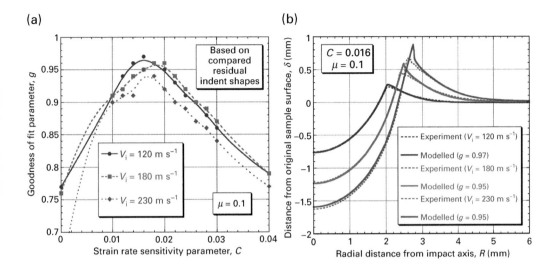

Fig. 10.25 Data [77] from residual indent prsofiles after impacts with different velocities, for samples of as-received Cu, showing (a) dependence of the goodness of fit parameter on the value of C and (b) comparisons between measured and modeled data, for the best-fit value of C.

Appendix 10.1 Calculating the Second Moment of Area, *I*

The second moment of area for a rectangular section beam of width w and thickness h is given by

$$I = \int_{-h/2}^{h/2} y^2 dA = \int_{-h/2}^{h/2} y^2 w\, dy = w\left[\frac{y^3}{3}\right]_{-h/2}^{h/2} = \frac{wh^3}{12} \qquad (10.23)$$

The corresponding operation for a circular cross-section of diameter D gives

$$I = \int y^2 dA = \int_{\theta=0}^{2\pi}\int_{r=0}^{D/2} (r\sin\theta)^2 [(rd\theta)dr] = \int_{\theta=0}^{2\pi}\left\{\int_{r=0}^{D/2} r^3 dr\right\} \sin^2\theta\, d\theta$$

$$= \frac{D^4}{64}\int_{\theta=0}^{2\pi} \sin^2\theta\, d\theta = \frac{D^4}{64}\int_{\theta=0}^{2\pi} \left(\frac{1-\cos 2\theta}{2}\right) d\theta = \frac{D^4}{64}\left[\frac{\theta}{2} - \frac{\sin 2\theta}{4}\right]_0^{2\pi}$$

$$\therefore I = \frac{\pi D^4}{64}. \qquad (10.24)$$

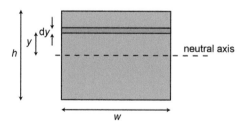

Fig. 10.26 Calculation of *I* for a rectangular section beam.

Appendix 10.1 Calculating the Second Moment of Area

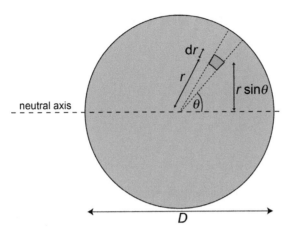

Fig. 10.27 Calculation of *I* for a circular section beam.

Appendix 10.2 Beam Deflections from Applied Bending Moments

As illustrated in Fig. 10.28, the beam curvature, κ, is approximately equal to the second derivative (curvature) of the neutral axis line (the dotted line in the diagram):

$$\kappa = \frac{d^2 y}{dx^2} \tag{10.25}$$

It can be seen that

$$M = \kappa EI = EI \frac{d^2 y}{dx^2} \tag{10.26}$$

Since the moment at the section concerned can also be written, for a cantilever beam, as

$$M = F(L - x) \tag{10.26}$$

it follows that

$$EI \frac{d^2 y}{dx^2} = F(L - x) \tag{10.27}$$

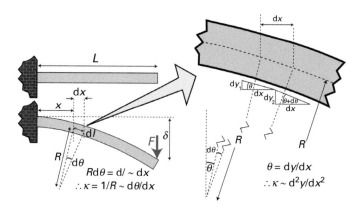

Fig. 10.28 The approximation involved in equating beam curvature to the curvature of the neutral axis of a beam.

Appendix 10.2 Beam Deflections from Bending Moments

This second-order linear differential equation can be integrated (twice), with appropriate boundary conditions, to find the deflection of the beam at different points along its length. For a cantilever beam, this operation is as follows:

$$EI\frac{dy}{dx} = FLx - \frac{Fx^2}{2} + C_1$$

$$\text{at } x = 0, \frac{dy}{dx} = 0, \therefore C_1 = 0$$

$$EIy = \frac{FLx^2}{2} - \frac{Fx^3}{6} + C_2$$

$$\text{at } x = 0, y = 0, \therefore C_2 = 0$$

This can be rearranged to give

$$y = \frac{Fx^2}{6EI}(3L - x) \tag{10.28}$$

For example, at the loaded end ($x = L$), this leads to

$$\delta = \frac{FL^3}{3EI} \tag{10.29}$$

For symmetrical three-point bending (Fig. 10.29), the bending moment is given by

$$M = \frac{Fx}{2}$$

It follows that

$$EI\frac{d^2y}{dx^2} = \frac{Fx}{2} \tag{10.30}$$

and the integration procedures lead to

$$EI\frac{dy}{dx} = \frac{Fx^2}{4} + C_1$$

$$\text{at } x = \frac{L}{2}, \frac{dy}{dx} = 0, \therefore C_1 = \frac{-FL^2}{16}$$

$$EIy = \frac{Fx^3}{12} - \frac{FxL^2}{16} + C_2$$

$$\text{at } x = 0, y = 0, \therefore C_2 = 0$$

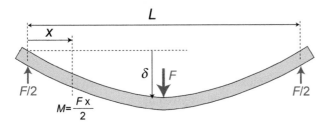

Fig. 10.29 Schematic diagram showing the central deflection for symmetrical three-point bending.

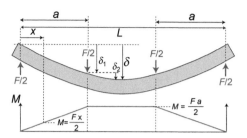

Fig. 10.30 Schematic diagram showing the central deflection for symmetrical four-point bending.

so the equation for the deflection is

$$y = \frac{-Fx}{48EI}\left(3L^2 - 4x^2\right) \qquad (10.31)$$

and the deflection at the center of the beam ($x = L/2$) is given by

$$\delta = \frac{-FL^3}{48EI} \qquad (10.32)$$

The other main loading configuration is symmetrical four-point bending (Fig. 10.30). In this case, the bending moment is given by

$$M = \frac{Fx}{2}$$

for $x \leq a$ and $x \geq (L-a)$, while

$$M = \frac{Fa}{2}$$

for $a \leq x \leq (L-a)$.

This derivation is slightly more complicated than that for three-point bending, since the value of dy/dx is not explicitly defined anywhere in the range $0 \geq x \geq a$, making it less simple to find the integration constant C_1. However, the fact that the curvature d^2y/dx^2 is defined (and constant) over the central region ($a \geq x \geq (L-a)$) and dy/dx = 0 at $x = L/2$, allows dy/dx to be found at $x = a$. The contribution to the overall deflection coming from the ($0 \geq x \geq a$) part of the beam, δ_1, is therefore obtained as follows. The initial integration is the same as for the three-point case:

$$EI\frac{dy}{dx} = \frac{Fx^2}{4} + C_1$$

Now, between the inner loading points, where the moment is constant ($M = Fa/2$), the curvature, which is also constant, is given by

$$\frac{d^2y}{dx^2} = \frac{M}{EI} = \frac{Fa}{2EI}$$

Appendix 10.2 Beam Deflections from Bending Moments

The change in gradient over a distance Δx in this regime can therefore be written:

$$\Delta\left(\frac{dy}{dx}\right) = \Delta x \left(\frac{d^2 y}{dx^2}\right)$$

Setting Δx equal to the distance between the center ($x = L/2$) and $x = a$, and recognizing that $dy/dx = 0$ at the center, allows the gradient at $x = a$ to be written as:

$$\left|\frac{dy}{dx}\right|_{x=a} = \left(\frac{-Fa}{2EI}\right)\left(\frac{L}{2} - a\right)$$

This allows the integration constant C_1 to be found:

$$EI\left(\frac{-Fa}{2EI}\right)\left(\frac{L}{2} - a\right) = \frac{Fa^2}{4} + C_1$$

$$\therefore C_1 = \frac{-Fa}{4}(L - a)$$

The other integration constant C_2 is then found:

$$EIy = \frac{Fx^3}{12} - \frac{Fax}{4}(L - a) + C_2$$

at $x = 0, y = 0, \therefore C_2 = 0$

Evaluating y at $x = a$ then gives

$$\delta_1 = \frac{Fa^3}{12EI} - \frac{Fa^2}{4EI}(L - a) = \frac{Fa^2}{12EI}[a - 3(L - a)] \qquad (10.33)$$

$$\therefore \delta_1 = \frac{-Fa^2}{12EI}(3L - 4a)$$

Between the inner loading points, where the moment is constant, the integration procedure is as follows, noting that the $x = 0$ point is now reset to $x = a$, with y reset to 0 there:

$$EI\frac{d^2 y}{dx^2} = \frac{Fa}{2}$$

$$EI\frac{dy}{dx} = \frac{Fax}{2} + C_3$$

at $x = \left(\frac{L}{2} - a\right), \frac{dy}{dx} = 0, \therefore C_3 = \frac{-FaL}{4} + \frac{Fa^2}{2}$

The final integration gives

$$EIy = \frac{Fax^2}{4} - \frac{FaLx}{4} + \frac{Fxa^2}{2} + C_4$$

at $x = 0, y = 0, \therefore C_4 = 0$

so the contribution to the deflection, δ_2, is

$$\delta_2 = \frac{1}{EI}\left[\frac{Fa}{4}\left(\frac{L}{2}-a\right)^2 - \frac{FaL}{4}\left(\frac{L}{2}-a\right) + \frac{Fa^2}{2}\left(\frac{L}{2}-a\right)\right]$$

(10.34)

$$\therefore \delta_2 = \frac{-Fa}{16EI}(L^2 - 4aL + 4a^2) = \frac{-Fa}{16EI}(L-2a)^2$$

The net deflection at the center of the beam ($x = L/2$) is thus given by

$$\delta = \delta_1 + \delta_2 = \frac{-Fa^2}{12EI}(3L - 4a) - \frac{Fa^2}{16EI}(L-2a)^2$$

$$\therefore \delta = \frac{-Fa}{48EI}(3L^2 - 4a^2)$$

(10.35)

The contribution from δ_2 only starts to become significant when L is greater than about $3a$. It may be noted that Eqn. (10.35) could actually be obtained by taking the expression for the deflection in three-point loading, Eqn. (10.31), and substituting $x = a$: this is, however, something of a happenstance, rather than a logical procedure for obtaining the expression.

Appendix 10.3 Mechanics of Springs

When a spring is loaded (compressed or extended), the deformation experienced by the wire is one of pure torsion. This is illustrated in Fig. 10.31.

The torque acting on the wire is given by

$$T = F\left(\frac{D}{2}\right)$$

in which F is the axial force and D is the coil diameter. Consider the small elemental length of the wire shown in Fig. 10.31, subtending an angle $d\beta$ at the axis of the coil. The torsional shear stress within the wire, τ, can be found by noting that it varies linearly with distance from the center of the wire ($\tau = Kr$, where K is an unknown constant and r is the distance of the element from the center of the wire). The torque can now be expressed in terms of internal forces in the wire. The force in an individual element of the wire is given by the torsional shear stress, τ, multiplied by the area of the element, $2\pi r\, dr$. Therefore, the elemental torque is simply this force multiplied by

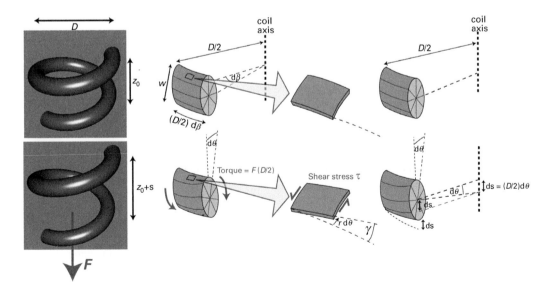

Fig. 10.31 Illustration of how the application of an axial load, F, to a spring generates torsional deformation of the wire and hence axial extension of the spring.

the distance of the element from the neutral axis of the wire, r. Summation of these elemental torques gives the total torque:

$$T = \int_0^{w/2} \tau(2\pi r \, dr) r = 2\pi K \int_0^{w/2} r^3 dr = \frac{\pi}{32} K w^4 \quad (10.36)$$

$$\therefore \tau = Kr = \frac{32Tr}{\pi w^4} = \frac{Tr}{2I}$$

in which I is the bending second moment of area (NOT the polar moment) for a solid cylinder (the wire), which is given (Appendix 10.1) by

$$I = \frac{\pi w^4}{64}$$

The local shear strain in the material is given by

$$\gamma = r \frac{d\theta}{dL} = \frac{r \, d\theta}{(D/2) \, d\beta}$$

in which $d\theta/dL$ is the rate of twist along the length of the wire. The incremental angle of twist of the wire is therefore related to the corresponding incremental rotation angle of the coil by

$$d\theta = \gamma \frac{(D/2)}{r} d\beta$$

The axial displacement of the coil, due to the twisting of the elemental section, can therefore be written

$$ds = \left(\frac{D}{2}\right) d\theta = \left(\frac{D}{2}\right) \gamma \frac{(D/2)}{r} d\beta$$

The axial displacement associated with the twist in one complete turn of the coil is thus

$$s = \left(\frac{D^2}{4r}\right) \gamma \int_0^{2\pi} d\beta = \left(\frac{\pi D^2}{2r}\right) \gamma$$

The local shear strain in the wire is therefore related to the increase in spacing between adjacent turns in the coil by

$$\gamma = \frac{2sr}{\pi D^2} \quad (10.37)$$

Measurement of the extension (per turn) of a spring, as a function of the applied force (first carried out systematically by **Robert Hooke**, in his pioneering work on the nature of elasticity) is a very convenient method of obtaining elastic constants. The ratio of τ to γ, obtained from the above equations, gives the shear modulus, G. This in turn can be related to the Young's modulus – see §2.4.4. The loading geometry is such that a large axial extension (per turn) is generated, while the strains within the material remain low, particularly for springs with a large ratio of D to w. Of course, this is exactly why springs are of practical use – they accommodate large deflections or displacements without the material being strained beyond its elastic limit.

Appendix 10.4 Interpretation of Data from Strain Gauge Rosettes

Rather than using single strain gauges, it is common to combine three of them in a single unit, as shown in Fig. 10.32. This allows the full strain state (magnitude and orientation of the principal strains) to be obtained. This can be done using Mohr's circle, an example of which is shown in Fig. 10.32.

Mohr's circle is fully defined by the three readings and the angle between them, which is 60° in real space and so 120° in Mohr space. (Strain is handled using Mohr's circle in exactly the same way as stress – see §2.2.3.) The radius of the Mohr's circle is r. Some trigonometry is now used to obtain the solution. The center of the circle is at the mean of the three readings (= 0.287). This is fairly intuitive, but can be formally shown as follows:

$$\varepsilon_A + \varepsilon_B + \varepsilon_C = 3\varepsilon_{centre} + r\{\cos 2\phi + \cos(120 - 2\phi) - \cos(60 - 2\phi)\}$$

Using the relationships

$$\cos(120 - 2\phi) = \cos\{180 - (60 + 2\phi)\} = -\cos(60 + 2\phi)$$

$$\cos(60 + 2\phi) + \cos(60 - 2\phi) = 2\cos 60 \cos 2\phi = \cos 2\phi$$

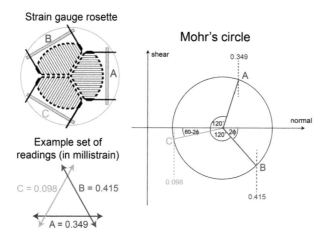

Fig. 10.32 Depiction of a (60°) strain gauge rosette and construction of a Mohr's circle to obtain the full strain state from an example set of readings.

It follows that

$$\varepsilon_{\text{centre}} = \frac{\varepsilon_A + \varepsilon_B + \varepsilon_C}{3}$$

In the example shown, it can be seen that

$$\cos 2\phi = \frac{\varepsilon_B - \varepsilon_{\text{centre}}}{r} = \frac{0.415 - 0.287}{r} = \frac{0.128}{r}$$

$$\cos(60 - 2\phi) = \frac{\varepsilon_{\text{centre}} - \varepsilon_C}{r} = \frac{0.287 - 0.098}{r} = \frac{0.189}{r}$$

$$\therefore \cos 2\phi = \frac{0.128}{0.189} \cos(60 - 2\phi) = 0.677(\cos 60 \cos 2\phi + \sin 60 \sin 2\phi)$$

This can now be solved to give the values of ϕ and r. Dividing through by $\cos 2\phi$,

$$1 = 0.677 \left(\frac{1}{2} + \frac{\sqrt{3}}{2} \tan 2\phi \right)$$

$$\therefore \tan 2\phi = \left(\frac{2}{0.677} - 1 \right) \frac{1}{\sqrt{3}} = 1.128 \rightarrow 2\phi = 48.4°$$

$$\therefore \phi = 24.2° \quad \text{and} \quad r = 0.193$$

In this example, the principal strains (in millistrain) are therefore given by

$$\varepsilon_x = 0.287 + 0.193 = 0.480$$

$$\varepsilon_y = 0.287 - 0.193 = 0.094$$

and the principal axes (of strain) lie at angles of 24.2° (x) and 65.8° (y) to the alignment direction of the B-gauge.

References

1. Jin, L, J Dong, J Sun and AA Luo, In-situ investigation on the microstructure evolution and plasticity of two magnesium alloys during three-point bending. *International Journal of Plasticity*, 2015. **72**: 218–232.
2. Dong, XL, HW Zhao, L Zhang, HB Cheng and J Gao, Geometry effects in four-point bending test for thin sheet studied by finite element simulation. *Materials Transactions*, 2016. **57**(3): 335–343.
3. Wang, T, OS Hopperstad, OG Lademo and PK Larsen, Finite element modelling of welded aluminium members subjected to four-point bending. *Thin-Walled Structures*, 2007. **45**(3): 307–320.
4. Lach, E and K Pohlandt, Testing the plastic behaviour of metals by torsion of solid and tubular specimens. *Journal of Mechanical Working Technology*, 1984. **9**(1): 67–80.
5. Toro, SA, PM Aranda, CM Garcia-Herrera and DJ Celentano, Analysis of the elastoplastic response in the torsion test applied to a cylindrical sample. *Materials*, 2019. **12**(19).

6. Bressan, JD and K Lopez, New constitutive equation for plasticity in high speed torsion tests of metals. *International Journal of Material Forming*, 2008. **1**: 213–216.
7. Khoddam, S and PD Hodgson, Post processing of the hot torsion test results using a multi-dimensional modelling approach. *Materials & Design*, 2010. **31**(5): 2578–2584.
8. Khoddam, S, PD Hodgson and MJ Bahramabadi, An inverse thermal-mechanical analysis of the hot torsion test for calibrating the constitutive parameters. *Materials & Design*, 2011. **32**(4): 1903–1909.
9. Zhou, M and MP Clode, A finite element analysis for the least temperature rise in a hot torsion test specimen. *Finite Elements in Analysis and Design*, 1998. **31**(1): 1–14.
10. Semiatin, SL, DW Mahaffey, NC Levkulich and ON Senkov, The radial temperature gradient in the Gleeble® hot-torsion test and its effect on the interpretation of plastic-flow behavior. *Metallurgical and Materials Transactions A: Physical Metallurgy and Materials Science*, 2017. **48A**(11): 5357–5367.
11. Clyne, TW and D Hull, *An Introduction to Composite Materials*, 3rd ed. Cambridge: Cambridge University Press, 2019.
12. McMeeking, RM, The finite strain tension torsion test of a thin-walled tube of elastic–plastic material. *International Journal of Solids and Structures*, 1982. **18**(3): 199–204.
13. Fedele, R, M Filippini and G Maier, Constitutive model calibration for railway wheel steel through tension-torsion tests. *Computers & Structures*, 2005. **83**(12–13): 1005–1020.
14. Lorenz, H, Remarks on the Euler buckling theory. *Zeitschrift Des Vereines Deutscher Ingenieure*, 1908. **52**: 827–831.
15. Born, M, A generalisation of Euler's buckling formula. *Physikalische Zeitschrift*, 1909. **10**: 383–390.
16. Zhu, XW, YB Wang and HH Dai, Buckling analysis of Euler–Bernoulli beams using Eringen's two-phase nonlocal model. *International Journal of Engineering Science*, 2017. **116**: 130–140.
17. Bardi, FC and S Kyriakides, Plastic buckling of circular tubes under axial compression – part I: experiments. *International Journal of Mechanical Sciences*, 2006. **48**(8): 830–841.
18. Tvergaard, V, Plastic buckling of axially compressed circular cylindrical shells. *Thin-Walled Structures*, 1983. **1**(2): 139–163.
19. Guillow, SR, G Lu and RH Grzebieta, Quasi-static axial compression of thin-walled circular aluminium tubes. *International Journal of Mechanical Sciences*, 2001. **43**(9): 2103–2123.
20. Legendre, J, P Le Grognec, C Doudard and S Moyne, Analytical, numerical and experimental study of the plastic buckling behavior of thick cylindrical tubes under axial compression. *International Journal of Mechanical Sciences*, 2019. 156: 494–505.
21. Reid, SR, Plastic deformation mechanisms in axially compressed metal tubes used as impact energy absorbers. *International Journal of Mechanical Sciences*, 1993. **35**(12): 1035–1052.
22. Lee, S, CS Hahn, M Rhee and JE Oh, Effect of triggering on the energy absorption capacity of axially compressed aluminum tubes. *Materials & Design*, 1999. **20**(1): 31–40.
23. Marzbanrad, J, A Keshavarzi and FH Aboutalebi, Influence of elastic and plastic support on the energy absorption of the extruded aluminium tube using ductile failure criterion. *International Journal of Crashworthiness*, 2014. **19**(2): 172–181.
24. Brazier, LG, On the flexure of thin cylindrical shells and other "thin" sections. *Proceedings of the Royal Society of London Series A: Containing Papers of a Mathematical and Physical Character*, 1927. **116**(773): 104–114.

25. Rotter, JM, AJ Sadowski and L Chen, Nonlinear stability of thin elastic cylinders of different length under global bending. *International Journal of Solids and Structures*, 2014. **51**(15–16): 2826–2839.
26. Karamanos, SA, Bending instabilities of elastic tubes. *International Journal of Solids and Structures*, 2002. **39**(8): 2059–2085.
27. Houliara, S and SA Karamanos, Stability of long transversely-isotropic elastic cylindrical shells under bending. *International Journal of Solids and Structures*, 2010. **47**(1): 10–24.
28. Roessle, ML and A Fatemi, Strain-controlled fatigue properties of steels and some simple approximations. *International Journal of Fatigue*, 2000. **22**(6): 495–511.
29. Ahlstrom, J and B Karlsson, Fatigue behaviour of rail steel – a comparison between strain and stress controlled loading. *Wear*, 2005. **258**(7–8): 1187–1193.
30. Adinoyi, MJ, N Merah and J Albinmousa, Strain-controlled fatigue and fracture of AiSi 410 stainless steel. *Engineering Failure Analysis*, 2019. **106**.
31. Griffith, AA, The phenomena of rupture and flow in solids. *Philosophical Transactions of the Royal Society*, 1920. **A 221**: 163–198.
32. Irwin, GR, Fracture dynamics, in *Fracturing of Metals*. ASM, 1948, pp. 147–166.
33. Paris, P and F Erdogan, A critical analysis of crack propagation laws. *Journal of Basic Engineering*, 1963. **85**: 528–534.
34. Watt, DF, A mechanism for production of intrusions and extrusions during fatigue. *Philosophical Magazine*, 1966. **14**(127): 87–92.
35. Karuskevich, M, O Karuskevich, T Maslak and S Schepak, Extrusion/intrusion structures as quantitative indicators of accumulated fatigue damage. *International Journal of Fatigue*, 2012. **39**: 116–121.
36. Polak, J, V Mazanova, M Heczko, R Petras, I Kubena, L Casalena and J Man, The role of extrusions and intrusions in fatigue crack initiation. *Engineering Fracture Mechanics*, 2017. **185**: 46–60.
37. Haghshenas, A and MM Khonsari, On the removal of extrusions and intrusions via repolishing to improve metal fatigue life. *Theoretical and Applied Fracture Mechanics*, 2019. **103**.
38. Ochi, Y, K Masaki, T Matsumura and T Sekino, Effect of shot–peening treatment on high cycle fatigue property of ductile cast iron. *International Journal of Fatigue*, 2001. **23**(5): 441–448.
39. Torres, MAS and HJC Voorwald, An evaluation of shot peening, residual stress and stress relaxation on the fatigue life of AiSi 4340 steel. *International Journal of Fatigue*, 2002. **24**(8): 877–886.
40. Gao, YK, Improvement of fatigue property in 7050–T7451 aluminum alloy by laser peening and shot peening. *Materials Science and Engineering: A Structural Materials: Properties, Microstructure and Processing*, 2011. **528**(10–11): 3823–3828.
41. Zhao, XH, HY Zhou, and Y Liu, Effect of shot peening on the fatigue properties of nickel-based superalloy GH4169 at high temperature. *Results in Physics*, 2018. **11**: 452–460.
42. Bag, A, D Delbergue, J Ajaja, P Bocher, M Levesque, and M Brochu, Effect of different shot peening conditions on the fatigue life of 300 M steel submitted to high stress amplitudes. *International Journal of Fatigue*, 2020. **130**.
43. Martin, V, J Vazquez, C Navarro and J Dominguez, Effect of shot peening residual stresses and surface roughness on fretting fatigue strength of Al 7075–T651. *Tribology International*, 2020. **142**.

44. Hussain, K, A Tauqir, A ul Haq and KQ Khan, Influence of gas nitriding on fatigue resistance of maraging steel. *International Journal of Fatigue*, 1999. **21**(2): 163–168.
45. Genel, K and M Demirkol, Effect of case depth on fatigue performance of AiSi 8620 carburized steel. *International Journal of Fatigue*, 1999. **21**(2): 207–212.
46. Tokaji, K, K Kohyama and M Akita, Fatigue behaviour and fracture mechanism of a 316 stainless steel hardened by carburizing. *International Journal of Fatigue*, 2004. **26**(5): 543–551.
47. Sirin, SY, K Sirin and E Kaluc, Effect of the ion nitriding surface hardening process on fatigue behavior of AiSi 4340 steel. *Materials Characterization*, 2008. **59**(4): 351–358.
48. Xiao, N, WJ Hui, YJ Zhang, XL Zhao, Y Chen and H Dong, High cycle fatigue behavior of a low carbon alloy steel: the influence of vacuum carburizing treatment. *Engineering Failure Analysis*, 2020. **109**.
49. Zhang, JM, SX Li, ZG Yang, GY Li, WJ Hui and YQ Weng, Influence of inclusion size on fatigue behavior of high strength steels in the gigacycle fatigue regime. *International Journal of Fatigue*, 2007. **29**(4): 765–771.
50. Krewerth, D, T Lippmann, A Weidner and H Biermann, Influence of non-metallic inclusions on fatigue life in the very high cycle fatigue regime. *International Journal of Fatigue*, 2016. **84**: 40–52.
51. Karr, U, B Schonbauer, M Fitzka, E Tamura, Y Sandaiji, S Murakami and H Mayer, Inclusion initiated fracture under cyclic torsion very high cycle fatigue at different load ratios. *International Journal of Fatigue*, 2019. **122**: 199–207.
52. D'Antuono, P, An analytical relation between the Weibull and Basquin laws for smooth and notched specimens and application to constant amplitude fatigue. *Fatigue & Fracture of Engineering Materials & Structures*, 2019. **43**(5): 991–1004.
53. Tanaka, K and Y Akiniwa, Fatigue crack propagation behaviour derived from *S-N* data in very high cycle regime. *Fatigue & Fracture of Engineering Materials & Structures*, 2002. **25**(8–9): 775–784.
54. Chapetti, MD, T Tagawa and T Miyata, Ultra-long cycle fatigue of high-strength carbon steels part I: review and analysis of the mechanism of failure. *Materials Science and Engineering: A Structural Materials: Properties, Microstructure and Processing*, 2003. **356**(1–2): 227–235.
55. Liu, YB, YD Li, SX Li, ZG Yang, SM Chen, WJ Hui and YQ Weng, Prediction of the S-N curves of high-strength steels in the very high cycle fatigue regime. *International Journal of Fatigue*, 2010. **32**(8): 1351–1357.
56. Nicholas, T, Tensile testing of materials at high rates of strain. *Experimental Mechanics*, 1981. **21**(5): 177–185.
57. Ma, DF, DN Chen, SX Wu, HR Wang, YJ Hou and CY Cai, An interrupted tensile testing at high strain rates for pure copper bars. *Journal of Applied Physics*, 2010. **108**(11).
58. Ben-David, E, T Tepper-Faran, D Rittel and D Shilo, A new methodology for uniaxial tensile testing of free-standing thin films at high strain-rates. *Experimental Mechanics*, 2014. **54**(9): 1687–1696.
59. Huh, H, JH Lim and SH Park, High speed tensile test of steel sheets for the stress–strain curve at the intermediate strain rate. *International Journal of Automotive Technology*, 2009. **10**(2): 195–204.
60. Sudheera, YS Rammohan and MS Pradeep, Split Hopkinson pressure bar apparatus for compression testing: a review. *Materials Today: Proceedings*, 2018. **5**(1): 2824–2829.

61. Lifshitz, JM and H Leber, Data processing in the split Hopkinson pressure bar tests. *International Journal of Impact Engineering*, 1994. **15**(6): 723–733.
62. Hartley, RS, TJ Cloete and GN Nurick, An experimental assessment of friction effects in the split Hopkinson pressure bar using the ring compression test. *International Journal of Impact Engineering*, 2007. **34**(10): 1705–1728.
63. Nie, HL, T Suo, BB Wu, YL Li and H Zhao, A versatile split Hopkinson pressure bar using electromagnetic loading. *International Journal of Impact Engineering*, 2018. **116**: 94–104.
64. Kariem, MA, RC Santiago, R Govender, DW Shu, D Ruan, G Nurick, M Alves, GX Lu and GS Langdon, Round-robin test of split Hopkinson pressure bar. *International Journal of Impact Engineering*, 2019. **126**: 62–75.
65. Xue, Q, LT Shen and YL Bai, A modified split torsional Hopkinson bar in studying shear localization. *Measurement Science and Technology*, 1995. **6**(11): 1557–1565.
66. Gilat, A and CS Cheng, Torsional split Hopkinson bar tests at strain rates above 10(4) S(−1). *Experimental Mechanics*, 2000. **40**(1): 54–59.
67. Yu, X, L Chen, Q Fang, XQ Jiang and YK Zhou, A review of the torsional split Hopkinson bar. *Advances in Civil Engineering*, 2018.
68. Holt, WH, W Mock, FJ Zerilli and JB Clark, Experimental and computational study of the impact deformation of titanium Taylor cylinder specimens. *Mechanics of Materials*, 1994. **17**(2–3): 195–201.
69. Maudlin, PJ, JF Bingert, JW House and SR Chen, On the modeling of the Taylor cylinder impact test for orthotropic textured materials: experiments and simulations. *International Journal of Plasticity*, 1999. **15**(2): 139–166.
70. Maudlin, PJ, GT Gray, CM Cady and GC Kaschner, High-rate material modelling and validation using the Taylor cylinder impact test. *Philosophical Transactions of the Royal Society A: Mathematical, Physical and Engineering Sciences*, 1999. **357**(1756): 1707–1729.
71. Lopatnikov, SL, BA Gama, MJ Haque, C Krauthauser, JW Gillespie, M Guden and IW Hall, Dynamics of metal foam deformation during Taylor cylinder–Hopkinson bar impact experiment. *Composite Structures*, 2003. **61**(1–2): 61–71.
72. Noh, HG, WJ An, HG Park, BS Kang and J Kim, Verification of dynamic flow stress obtained using split Hopkinson pressure test bar with high-speed forming process. *International Journal of Advanced Manufacturing Technology*, 2017. **91**(1–4): 629–640.
73. Mylonas, GI and GN Labeas, Mechanical characterisation of aluminium alloy 7449–T7651 at high strain rates and elevated temperatures using split Hopkinson bar testing. *Experimental Techniques*, 2014. **38**(2): 26–34.
74. Sedighi, M, M Khandaei and H Shokrollahi, An approach in parametric identification of high strain rate constitutive model using Hopkinson pressure bar test results. *Materials Science and Engineering: A Structural Materials: Properties, Microstructure and Processing*, 2010. **527**(15): 3521–3528.
75. Sasso, M, G Newaz and D Amodio, Material characterization at high strain rate by Hopkinson bar tests and finite element optimization. *Materials Science and Engineering: A Structural Materials: Properties, Microstructure and Processing*, 2008. **487**(1–2): 289–300.
76. Mishra, A, M Martin, NN Thadhani, BK Kad, EA Kenik and MA Meyers, High-strain-rate response of ultra-fine-grained copper. *Acta Materialia*, 2008. **56**(12): 2770–2783.
77. Burley, M, JE Campbell, J Dean and TW Clyne, Johnson–Cook parameter evaluation from ballistic impact data via iterative FEM modelling. *International Journal of Impact Engineering*, 2018. **112**: 180–192.

Index

acoustic emissions, 100
acoustic waves, 69
activation energy, 35, 75, 78, 203
age-hardening, 65
ageing, 64
Al, 55, 59, 130
Al_2Cu, 64
alloy
  Al, 99
  Al-Cu, 64
  Cu-Ni, 59
  Cu-Zn (brass), 70
  Fe-C, 71
  Ni superalloy, 63
  Ni-50Ti, 71
  Ni-Ti, 33
  shape memory, 115
  superelastic, 115
alumina, 195
aluminium powder, 66
American Society of Testing and Materials, 81
amorphous materials, 66
amorphous metals, 201
anisotropy, 37, 85, 100–101, 117, 139, 152, 162, 167, 169
annealing, 86
annealing twins, 70
Armco Iron, 62
Arrhenius, 34, 75, 78
ASTM, 81
austenite, 73, 99

back-stress, 117
ballistic impact, 74
ballistic indentation, 249
ballistics, 31
barreling, 108, 112, 113, 119, 159
Basquin law, 242
Bauschinger
  Johann, 117
Bauschinger effect, 107, 117
bcc structure, 60
beam bending, 219

beam curvature, 219, 256
beam deflections, 221, 256
beam stiffness, 221, 232
bedding down, 113
bend test fixtures, 225
bending moments, 219
Berkovich indenters, 137
biaxial tensile testing, 100
bifurcation buckling, 234
bonding
  covalent, 43
  ionic, 43
  metallic, 43
brass, 130
brasses, 99
Brazier buckling, 234
Brinell
  Johan August, 125
Brinell hardness test, 125, 134
British Standards Institute, 81
brittle materials
  bend testing, 225
BSI, 81
buckling, 108, 209
  Brazier, 234
  hollow tubes, 232
bulk modulus, 19
Burgers
  Johannes, 44
Burgers circuit, 46
Burgers vector, 46, 50, 55
butt-end shoulder, 84

C250 maraging steel, 155
cantilever beam, 219, 256
cantilever loading, 87
carburizing, 242
castings, 5, 40
cavitation, 97
cermet, 124, 163, 173, 196
characteristic strain, 29
chemical etching, 156
Chernov–Lüders band, 98

climb, 29, 55, 76
clip gauges, 88, 108, 113
close-packed directions, 46
close-packed planes, 46
coarsening, 65
coatings, 193, 201
Coble creep, 75
coefficient of friction, 160, 228
coefficient of restitution, 143
coherency strains, 65
coherent boundaries, 70
coherent boundary, 68
cold work, 57
compliance calibration, 87, 158
compliance tensor, 17
concertina folding, 233
conductivity, 61
   electrical, 43
   thermal, 43
Considère
   Armand, 89
Considère construction, 89, 94–95
constitutive law, 29, 127, 165
constitutive laws, 59, 73, 78, 115
constraint, 86, 207, 213
constraint effect, 58
constriction, 55
contact area, 125
contact drift, 198
contact stress, 225
convergence, 149
convergence algorithms, 164
convergence criterion, 165
copper, 61, 94, 139, 249
Cottrell
   Alan, 60
Cottrell atmospheres, 60, 62, 98
crack initiation, 242
crack opening displacement, 241
crack propagation, 3
crack tip
   plastic zone, 241
crack tip opening displacements, 88
cracking, 139
crashworthiness, 234
creep, 1, 34, 67, 74, 102, 124
   Coble, 75
   diffusional, 76
   dislocation, 76
   from nanoindentation data, 203
   Nabarro–Herring, 76
   primary, 34, 173
   secondary, 34
   tertiary, 34, 176
critical failure strain, 98
critical fracture strain, 239
critical strain criterion, 30

cross-head, 87
cross-slip, 29, 54
cruciform samples, 100
crystal structure, 85, 100
   hexagonal, 69–70
   tetragonal, 69
crystallographic texture, 58
Cu, 55, 86, 110, 130
cut through, 66
cutting, 31
cyclic loading, 3, 87, 235
cycling frequency, 242

dead weight loading, 87, 102
deep drawing, 39
deep-drawn sheet, 101
deformation twinning, 25, 32, 67, 74, 159
delocalized electrons, 43
depth-sensing indentation, 192
determinant, 11
diagonalising, 11
diamond, 124, 195
   oxidation, 199
diamond cone, 129
diamond indenters
   oxidation, 137
diamond tips, 137
DIC (digital image correlation), 59, 85, 88
diffusion, 74, 78
diffusion distances, 61
diffusional creep, 76
dilation, 16
direction cosines, 10–11, 13
dislocation, 4, 21, 29, 43
   annihilation, 56
   climb, 53, 55
   creep, 76
   cross-slip, 54, 77
   density, 51, 53, 57, 77, 85, 193, 206
   edge, 45, 60
   energy, 52
   energy of, 48
   force on, 48, 64
   forces between, 52
   forest hardening, 206
   geometrically necessary, 206
   glide, 21, 25, 44, 48, 67, 101, 159
   glissile, 53
   jogs, 22, 206
   kinks, 53
   line tension, 49, 64
   line vector, 46
   mixed, 47
   mobility, 29, 117, 193
   partial, 48, 55
   perfect, 48, 55
   screw, 44, 47

Index

dislocation (cont.)
  sessile, 53
  spacing between, 52
  strain field, 52
  tangles, 22, 29, 118, 206
dispersion strengthening, 63, 66
displacement control, 163
displacement measurement, 87
distribution of plastic strain, 153
DoITPoMS, 46, 90, 221
DSC (differential scanning calorimetry), 33
dual phase steels, 116
ductile rupture, 239
ductility, 29, 61, 92
dummy suffix, 10

earing, 101
easy glide, 99
eddy current gauges, 88, 108
EDM (electro-discharge machining), 82
eigenvalues, 11
einstein summation convention, 10, 17
elastic constants, 17
elastic strain energy, 24
electro-magnetic induction, 143
electron beam melting, 86
electron microscopes, 85
elongation at failure, 92, 98
engineering shear strain, 18
Euler
  Leonard, 231
Euler buckling, 231
Euler buckling stress, 109
explosions, 31, 70
extrusion, 66

face-centred cubic, 46
failure strain, 92
fatigue, 235, 241
  high cycle, 236
  low cycle, 236
fatigue life, 242
fatigue limit, 242–243
fatigue resistance, 37
fcc, 46, 55, 68, 71
Fe-C martensite, 71
FEM (finite element method), 3, 29–31, 40, 92, 97,
  112, 115, 119, 127, 142, 150, 155, 158, 166,
  170, 196, 226
  inverse, 40, 148
  iterative, 173
  mesh independence, 160
  numerical stability, 160
  sensitivity analyses, 159
ferrite, 61, 116
FIB (focussed ion beam), 209
fibre composites, 230

fibre push-out, 209
field tensors, 16
first rank tensor, 7
flow stress, 28–29, 134, 144, 152
force balance, 37, 118
forest hardening, 53
forging, 37, 39, 56, 89, 169
four-point bend testing, 226, 258
fracture, 3
  critical plastic strain, 166
  critical strain, 3
fracture criterion, 93
fracture energy, 3, 238
fracture mechanics, 3, 236
fracture strength, 238
fracture toughness, 3, 82, 88, 193, 240
frame drift, 198
Frank
  Charles, 44
Frank's rule, 49, 55
Frank–Read source, 54, 206
free suffix, 10
Frenkel
  Jacov, 44
friction, 108, 110, 116, 119, 159
friction coefficient, 108, 112, 119
friction stir processing, 86
gas gun, 252
gauge length, 81, 85, 88, 92, 107

geometrically necessary dislocations, 206
glassy metals, 201
goodness-of-fit, 164
grain boundaries, 58, 75, 78, 85, 118
grain boundary sliding, 150
grain boundary structure, 139
grain rotations, 140
grain size, 58, 76, 85–86, 139, 142, 150
grain structure, 135, 138
Griffith fracture criterion, 236, 238
grinding, 86
grips, 81
  Hounsfield, 83
  pinned, 82
  serrated, 82
  split-collar, 82
  threaded, 82
growth twins, 70
Guinier–Preston zones, 64

Hadfield's manganese steel, 116
hardness testing, 2, 58, 85, 87, 118, 123–125, 204
  Brinell, 125
  conversion between hardness numbers, 131
  effect of residual stresses, 140
  from load–displacement data, 203
  Knoop, 137

Leeb, 143
Mohs, 144
Rockwell, 129, 140
Vickers, 129, 131
HCF (high cycle fatigue), 236
hcp, 71
heat treatment, 29, 65
heating from plastic deformation, 101
high cycle fatigue, 236
high strain rate tensile testing, 101, 245
high strain rates, 143
high-speed photography, 252
Hirsch
    Peter, 44
homologous temperatures, 35
Hooke
    Robert, 262
Hooke's law, 17
Hopkinson bar tests, 247
hot working, 58, 61
HY-100 steel, 93
hydraulic systems, 87
hydrostatic line, 23, 26
hysteresis, 33, 72

impact events, 69, 247
indent profiles, 140, 155, 158, 161
indentation, 1
    cryogenic, 198
    depth-sensing, 138, 192
    in vacuum, 199
    pop-in, 205
    size effect, 193
indentation creep
    pile-up, 177
    use of a recess, 173
indentation creep plastometry, 103, 173
indentation plastometry, 2, 85, 97, 115, 118, 124,
    127, 140, 149, 194, 212, 228, 250
    commercialization, 179
    curved surfaces, 156
    edge effects, 157
    friction, 160
    inclined surfaces, 157
    sample preparation, 154
indentation superelastic plastometry, 178
indenter shapes, 124
    self-similar, 124
indenters
    area function, 125, 202
    Berkovich, 137, 141, 196
    damage, 137, 196
    Knoop, 141
    oxidation, 196
    sharp, 125, 136, 196
    spherical, 149
    Vickers, 196

inhomogeneity, 167
Instron Corporation, 129
interfacial energies, 65
interfacial sliding, 162
interferometric optical set-ups, 88
internal oxidation, 61
interstitial
    carbon, 61
    nitrogen, 61
    oxygen, 61
interstitial solute, 60
intrusion–extrusion mechanism, 242
invariants, 12
inverse FEM, 97, 148
ion beam milling, 109
ion implantation, 210
iterative FEM, 119
iterative modeling, 40

jogs, 53
Johnson–Cook law, 31, 39, 244

Knoop
    Frederick, 137
Knoop test, 137

laser cutting, 82
laser peening, 242
lattice friction stress, 45, 59
lattice vector, 48
LCF (low cycle fatigue), 236
lead screws, 87
Leeb test, 143
lenticular, 70
linear work hardening, 27
load capability, 85
load cell, 87
load drop, 98
load–displacement plot, 179
loading frame, 87
loading rate, 5
loading train, 87, 158
low cycle fatigue, 236
lubrication, 107, 108, 110, 117, 119, 136, 161
Lüders bands, 62, 98
Ludwigson equation, 159
Ludwik–Hollomon equation, 29, 31, 62, 90, 157,
    159
LVDT (linear variable displacement transducers),
    88, 108

machining, 31
Mangalloy, 116, 128
maraging steel, 163
Martens
    Adolf, 71
martensitic transformations, 71, 73, 74, 99

matter tensors, 17
maximum stress criterion, 230
Maxwell
  James Clerk, 24
mechanical "backlash", 129
mechanical twins, 67
meso-scale, 150
metal forming, 56
metal matrix composites, 5
metallic foams, 6
metastable precipitates, 65
micrographs, 65
micropillar compression, 109, 193, 208
microstructure, 2, 43, 59, 77, 85, 116, 193, 242
microstructure, 77
mild steel, 61, 98–99
Miller indices, 23, 46, 51
Miller–Norton law, 35, 174
misfit strains, 37, 40
misorientation angles, 57
MMC (metal matrix composites), 5, 169
Mohr
  Christian Otto, 12
Mohr's circle, 12, 25, 248, 264
Mohs
  Friedrich, 123, 145
Mohs hardness scale, 144
moment balance, 37
$MoSi_2$, 108
multiple slip, 58

Nabarro–Herring creep, 76
nanoindentation, 2, 135, 168, 192
nanoindentations, 148, 152
  size effect, 204
necking, vii, 30, 89, 116, 151, 159, 238
Nelder–Mead convergence algorithm, 183
Nelder–Mead simplex search, 164
Neumann bands, 69
neutral axis, 256
Ni, 173
Ni-base superalloy, 139, 210
nitriding, 242
non-metallic inclusions, 242
normal stress, 8

octahedral interstice, 60
Orowan
  Egon, 44, 63
Orowan bowing, 63
Orowan stress, 64
over-ageing, 66
oxidation of diamond, 137
oxide films, 119, 130, 136, 150, 154, 207
oxygen in copper, 162
oxygen-free high conductivity Cu, 61

parallel plate capacitors, 88
Paris–Erdogan law, 241
Pb, 119
Peierls stress, 45, 59, 139
penetration ratio, 129, 151
persistent slip bands, 67, 99, 139, 212
phase
  austenitic, 33
  martensitic, 33
phase diagram
  Al-Cu, 64
phase transformations, 115, 207
  martensitic, 32
pile-ups, 118, 125, 132, 153, 162, 202
pin-jointed ends, 232
pinning, 61
pipeline, 156
plane strain, 240
plane stress, 26
plastic buckling, 232
plastic strain
  distribution, 153
plastic work
  distribution in strain level, 185
platens, 108, 119
Poisson ratio, 4, 18, 162, 202
Polanyi
  Michael, 44
polar second moment of area, 223
polishing, 86
polycrystals, 21, 58
polygonization, 56
polymers, 242
polytope, 165
pop-in, 206
porosity, 5
porous materials, 115
Portevin–Le Chatelier effect, 62, 98
precipitates, 59, 64, 77, 138, 201
precipitation hardening, 63, 66
pre-load, 130
premature fracture, 84
primary creep, 34
primary slip system, 50, 211
principal strains, 16, 95
principal stresses, 11
profilometers, 181
profilometry, 158
projected contact area, 204
pure shear, 248

quasi-static, 82
quenching, 64

RA (reduction of area), 89, 93–94
R_a value, 155
radiation, 29
radius of gyration, 232

Index

Ramberg–Osgood equation, 159
rate of twist, 262
Read
    Thornton, 44
rebound hardness, 142
recovery, 56
recrystallization, 36, 39, 57, 61, 77
reduced modulus, 202
reduced section length, 84
reduced sum of squares, 164
replica technique, 158
representative volume, 85, 126, 135, 150, 204
residual stresses, 36, 59, 117, 140, 169, 242
    equal biaxial, 140
resolved shear stress, 50
rigid body rotation, 14
ring compression test, 119
Rockwell
    Hugh, 129
    Stanley, 129
Rockwell hardness test, 129, 134
rolling, 37, 39, 56, 89
R_z value, 155

sample size, 82, 85, 109
sapphire, 195
scalars, 7
scanning laser extensometry, 88, 108
Schmid factor, 50, 58, 211
Schmid's law, 50
Schmidt hammer test, 142
scratch testing, 144
screw dislocation, 47
second moment of area, 220, 254
    circular section, 254
    rectangular section, 254
second rank tensors, 7, 11
secondary creep, 34
secular equation, 12
serrations, 61, 98
servo-hydraulic valves, 87
shape memory alloys, 32, 116
shear bands, 66
shear modulus, 18, 20, 44, 262
shear stress, 8
    resolved, 50
shock waves, 32
shore test, 143
shot peening, 37, 242
shoulders, 82
SHPB (split Hopkinson pressure bar), 101, 247
simple cubic structure, 45
simplex, 165
single crystal, 50, 138, 150, 168, 193
single crystals, 29, 44, 58, 76, 82, 99–100, 109, 117, 209

single crystals, 62
sink-in, 125, 132, 162
size effects, 136, 203, 204
slenderness ratio, 232
slip plane, 46, 48
slip system, 46
    primary, 50
slip systems, 21
SME (shape memory effects), 32, 33, 73
solid solubility, 59
solute atoms, 59
    interstitial, 60
    substitutional, 59
solution strengthening, 59
solution treatment, 64
speckles, 88
spherical indenters, 137
springs, 224, 261
stable necking, 92
stacking fault, 55
stacking fault energy, 55
stacking sequence, 71
stainless steel, 31, 233
steels, 61, 69
    AISI-1018, 249
    AISI-4340, 249
    dual phase, 116
    duplex stainless, 127
    hadfield's manganese, 116
    maraging, 248
    mild, 61
    SPFH590, 246
    SS400, 246
    stainless, 31
    TRIP, 99, 116
    X-80, 118
    08PS, 99
sticking friction, 108, 119
stiffness tensor, 17
strain energy release rate, 237
strain gauge rosettes, 248, 264
strain gauges, 88, 264
strain gradient plasticity, 206
strain hardening, 21
strain rate dependence of plasticity, 243
strain rate sensitivity, 32, 39, 144
strains, 2, 7
    average plastic, 129
    deviatoric, 16
    hydrostatic, 16, 60
    measurement, 84
    nominal, 88
    principal, 95, 248
    von Mises, 23, 95, 128, 159
stress concentration, 238
stress exponent, 36, 75, 173, 203
stress intensity factor, 240

stress ratio, 241
stress space, 23, 26
stresses, 2, 7
  bowing, 64
  concentration, 81
  contact, 225
  deviatoric, 22, 44, 115
  engineering, 27
  exponent, 78
  flow, 28, 134
  hydrostatic, 19, 22, 115, 159
  lattice friction, 45
  nominal, 27, 88
  normal, 8
  Orowan, 64
  Peierls, 45, 139
  principal, 11
  relaxation, 73
  residual, 36, 169
  saturation, 29
  shear, 8, 23, 44
  threshold, 241
  true, 27
  ultimate tensile, 91
  von Mises, 22, 115, 124, 134, 159, 251
stress-life fatigue testing, 241
stretcher strains, 98
sub-critical crack growth, 236, 239
sub-grain boundaries, 56
sub-grains, 56
substitutional solute, 59
superelasticity, 32, 72
supersaturated solid solution, 64
surface roughness, 119, 130, 136, 143, 150, 154, 201, 228
swaging, 251
symmetry, 17

Tabor
  David, 123
Taylor
  Geoffrey, 44
taylor cylinder test, 249
taylor factor, 58
temperature, 5
  homologous, 56
tensile drawing of polymers, 92
tensile testing, 1, 50, 81
tension-compression asymmetry, 100, 107, 115, 116, 159
tension-torsion tests, 230
tensorial shear strain, 18
tensors, 2
  anti-symmetrical, 14
  compliance, 17
  deformation, 14
  field, 16

  first rank, 7
  matter, 17
  relative displacement, 14
  rotation, 15
  second rank, 7
  stiffness, 17
  strain, 14
  symmetrical, 14
  zeroth rank, 7
testing standards, 81
texture, 58, 85, 101, 117, 139, 150, 152, 168
thermal activation, 56
thermal drift, 194, 198, 204
thin sheet, 82, 85
three-point bending, 257
tilt boundary, 56
tin cry, 69
torque, 221
torsion, 221, 261
torsion testing, 228
torsional coils, 224
torsional split Hopkinson bar, 248
toughness, 195
transformation of axes, 9
transformation-induced plasticity, 99, 116
transformations
  martensitic, 115
Tresca, 172
  Henri, 25
Tresca yield criterion, 25, 100
true stress, 27
Tsai–Hill criterion, 230
tungsten carbide, 130
twin
  boundaries, 68
twinning, 21
  deformation, 21
  direction, 67
  elements, 67
  plane, 67
  shear, 67
twist boundary, 57
twisting moment, 221

ultimate tensile stress, 3–4
uniqueness, 149
unit cell, 4, 46
untextured, 58
UTS (ultimate tensile stress), 3, 86, 92, 97, 116, 126, 151

vacancy, 55
vector, 7, 9
very soft materials, 156
Vickers equivalent cone, 133
Vickers hardness, 152, 192
  values, 132

Vickers testing machine, 131
Voce equation, 29, 31, 62, 93–94, 159
von Mises
   Richard Edler, 24
von Mises strain, 95
von Mises stress, 22
von Mises yield criterion, 101

Wade
   William, 123
water jet cutting, 82
welding, 40
wood, 232
work hardened layer, 156
work hardening, 3, 21, 89, 132, 143, 151, 162, 238
   linear, 27

work hardening coefficient, 27, 29
work hardening exponent, 29
work hardening rate, 28, 56, 116
worm drives, 87
wrinkling, 232

yield envelope, 23, 26
yield point, 58
yield stress, 3–4, 22, 140, 143
yielding criteria
   Tresca, 25
   von Mises, 24, 172
Young's modulus, 4, 17, 88, 138, 162, 168, 201–202, 220, 232, 263

π-plane, 26